ADVANCE PRAISE FOR

For the Love of Nature: Ecowriting the World

This engaging collection posits the value of and need for instruction on ecowriting that fosters students' relationships with the natural world and an appreciation of Indigenous/Non-Western ecological perspectives. The book includes useful essays on teaching various types and genres of ecowriting, including the use of digital media productions such as video and multi-modal essays. It also includes examples of students' essays, letters, and poems from Jeff Share's environmental justice class that illustrate the instructional methods described in the book. Therefore, this book is a useful resource for teachers to incorporate ecowriting into their teaching to engage students in creatively portraying their experiences with the natural world for having them address the need for action about the global climate crisis.

—Richard Beach, Professor Emeritus of English Education,
University of Minnesota

Drawing upon Indigenous wisdom and critical pedagogy, *For the Love of Nature: Ecowriting the World* is both a call to action and an important resource for teaching, learning, and enacting environmental justice. The beautiful collection of essays is a must-read for all K-12 educators interested in advancing educational and climate justice to create a more sustainable and caring world for our present and our future.

—Annamarie M. Francois, Ed.D. Executive Director Center X,
University of California, Los Angeles

For the Love of Nature

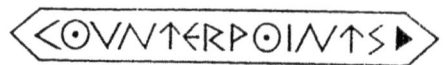

STUDIES IN CRITICALITY

Shirley R. Steinberg
Series Editor

Vol. 547

For the Love of Nature

Ecowriting the World

Edited by Jeff Share

New York · Berlin · Bruxelles · Chennai · Lausanne · Oxford

Library of Congress Cataloging-in-Publication Data

Names: Share, Jeff, author.
Title: For the love of nature : ecowriting the world / edited by Jeff Share.
Description: New York : Peter Lang, 2024. | Series: Counterpoints, 1058-1634 ; vol 547 | Includes bibliographical references.
Identifiers: LCCN 2023040619 (print) | LCCN 2023040620 (ebook) | ISBN 9781433199790 (paperback) | ISBN 9781433199806 (pdf) | ISBN 9781433199813 (epub)
Subjects: LCSH: Natural history—Authorship. | Ecology—Authorship. | Natural history literature—Study and teaching. | Ecoliterature—Study and teaching. | LCGFT: Essays.
Classification: LCC QH14 .F67 2024 (print) | LCC QH14 (ebook) | DDC 508—dc23/eng/20231106
LC record available at https://lccn.loc.gov/2023040619
LC ebook record available at https://lccn.loc.gov/2023040620
DOI 10.3726/b21296

Bibliographic information published by the Deutsche Nationalbibliothek.
The German National Library lists this publication in the German National Bibliography; detailed bibliographic data is available on the Internet at http://dnb.d-nb.de.

Cover design by Peter Lang Group AG

ISSN 1058-1634 (print)
ISBN 9781433199790 (paperback)
ISBN 9781433199806 (ebook)
ISBN 9781433199813 (epub)
DOI 10.3726/b21296

© 2024 Peter Lang Group AG, Lausanne
Published by Peter Lang Publishing Inc., New York, USA
info@peterlang.com - www.peterlang.com

All rights reserved.
All parts of this publication are protected by copyright.
Any utilization outside the strict limits of the copyright law, without the permission of the publisher, is forbidden and liable to prosecution.
This applies in particular to reproductions, translations, microfilming, and storage and processing in electronic retrieval systems.

This publication has been peer reviewed.

This book is dedicated to all who continue to believe in a better world and work for social and environmental justice, in spite of the odds and obstacles.

Contents

Epigraph	1
No Más Humanos/No More Humans	3
SARAWI ANDRANGO	
Land Acknowledgment	5
MANDIE TORRES	
Acknowledgments	9
Introduction	11
JEFF SHARE	

Part I: Exploring Ecowriting

1. *'If You Win the Popular Imagination, You Change the Game':
   Why We Need New Stories on Climate* — 23
   REBECCA SOLNIT

2. *Ecowriting: A Fieldguide* — 37
   GAVIN LAMB

3. *Let's Story about Storying* — 57
   DENISE CHAPMAN

4. *Beyond Nature's Children: Examining the Environmental,
   Cultural, and Political Influences that Inform Indigenous
   Perspectives and Stories about Nature and the Environment* — 67
   MELISSA GREENE-BLYE

Part II: Teaching Ecowriting

5. *Finding a Place in the World: Ecowriting with Elementary and Middle School Students* — 77
   CINDY JENSON-ELLIOTT

6. *Ecomedia Video Essays* — 85
   ANTONIO LÓPEZ

7. *Visual Journaling as Ecowriting* — 95
   PEACHES HASH AND THERESA REDMOND

8. *Disrupting Hierarchies of Power and Uplifting Environmental Justice through Collaborative Ecowriting* — 111
   ALEJANDRO OJEDA, ELMER ORTEGA, JENIFER RAMOS, VANESSA ROMERO, NEIDA SANDOVAL-LOPEZ, BENJAMIN THOMPSON, MARÍA VERÓNICA VALERIANO, AND JAZMINE VEGA LOPEZ

9. *Grand Appreciation for All Things Natural* — 117
   ROSE WHITE

Part III: Student Ecowriting

Poems:

10. *A Meaningful Purpose* — 121
    KATHY LIZAOLA

11. *The Life of a Fast Fashion Garment Worker* — 123
    GABRIELA VENEGAS

12. *Is It Even Air?* — 125
    GISELLE VILLANUEVA

13. *On Vacation* — 127
    NEIDA SANDOVAL-LOPEZ

14. *Hard to Appreciate* — 129
    KEIMORA NETTLES

15. *Feels Like Home* — 131
    ANTONIA BURGARD, LUDMILLA SEMSKOW, LEA IRINA HEUING, JANA KRANZ, AND PHILIPPA WITZENHAUSEN

Contents

Short Stories:

16. *Divisions That Destroy Us* Sara Fernandez	133
17. *Water* Rucha Deshpande	135
18. *Elena and the Mountain* Vanessa Romero	137
19. *The Evolution of the Apple* Elmer Ortega	141
20. *Gardening in the Time of Covid* Esmeralda Orozco Sanchez	143
21. *The Manifestation of Taoism in Environmental Protection* (Alice) Yanan Sun	145
22. *Armando* María Verónica Valeriano	147
23. *I Am a Drop of Water* Yaying Wu	153

Letters:

24. *Dear Diary* Arbrean Sears	155
25. *To Our Beautiful Mother* Mandie Torres	159
26. *Dear Councilwoman Monica Rodriguez* Nicole Hall	163

Part IV: Resources for Ecowriting

Lesson plans:

27. *Introduction to Ecowriting Lesson Units* Sydney Richmond and Andrea Gambino	167

28. Ecowriting Unit One: Exploring Our Relationships with Nature SYDNEY RICHMOND AND ANDREA GAMBINO	171
29. Ecowriting Unit Two: Greenwashing – Disrupting False or Misleading Claims of Environmental Ethics SYDNEY RICHMOND AND ANDREA GAMBINO	181
30. Ecowriting Unit Two: Greenwashing – Instructional Resources Guide SYDNEY RICHMOND AND ANDREA GAMBINO	201
31. Recommended Resources	207
Notes on Contributors	215
About the Editor	221

Epigraph

The story of our relationship to the earth is written more truthfully on the land than on the page. It lasts there. The land remembers what we said and what we did. Stories are among our most potent tools for restoring the land as well as our relationship to land. We need to unearth the old stories that live in a place and begin to create new ones, for we are storymakers, not just storytellers (Kimmerer, 2013, p. 341).

References

Kimmerer, R. W. (2013). *Braiding sweetgrass: Indigenous wisdom, scientific knowledge, and the teachings of plants*. Milkweed.

From *Braiding Sweetgrass* by Robin Wall Kimmerer (Minneapolis: Milkweed Editions, 2013). Copyright © 2013 by Robin Wall Kimmerer. Reprinted with permission from Milkweed Editions. milkweed.org

No Más Humanos/No More Humans

SARAWI ANDRANGO

NO MÁS HUMANOS,
Del virus humano nada bueno se espera
el ser esencial que está prisionero en él,
  debe ser liberado
para salvar la continuidad.
¡Despierta!
Si el humano, no escucha, no ve, no
  siente, no trasciende
¿para qué existe?
La soledad no existe,
caminas ciego más bien
mientras exista una hormiga, el viento,
  una flor;
nos acompañamos
solo que no eres capaz de verlos menos
  aún sentirlos.
¡Despierta!
Si el humano, no escucha, no ve, no
  siente, no trasciende
¿para qué existe?
Las religiones estorban
sus credos señalan al humano como
  superior
prometiendo el paraíso a costa de
  diezmos, sumisión, muerte
saquean la paz y armonía del corazón
hay que desprenderse del apego a toda
  figura externa.
¡Despierta!
Si el humano, no escucha, no ve, no
  siente, no trasciende
¿para qué existe?

NO MORE HUMANS,
From the human virus, nothing good
  will come
the essential being that is imprisoned in
  it, must be liberated
to save continuity.
Wake up!
If humans don't listen, don't see, don't
  feel, don't transcend
why exist?
Solitude does not exist,
rather, you walk blindly
as long as an ant, the wind, a flower
  exists;
we are accompanied
It is just that you are not able to see
  them, much less feel them.
Wake up!
If humans don't listen, don't see, don't
  feel, don't transcend
why exist?
Religions interfere
their creed position humans as superior
promising paradise at the cost of tithes,
  submission, death
pillaging peace and harmony from
  the heart
one must shed the attachment to all
  external forms.
Wake up!
If humans don't listen, don't see, don't
  feel, don't transcend
why exist?

El individualismo es la debilidad
la existencia armónica necesita de la
  dualidad
opuestos complementarios para dar
  equilibrio a la vida
la unión de estos extremos difumina
  vibraciones
El individualismo sobrevive, la dualidad
  co - existe.
¡Despierta!
Si el humano, no escucha, no ve, no
  siente, no trasciende
¿para qué existe?
La medicina eres tú
toda enfermedad es creada
usada como arma por las transnacionales
te envenenan – te curan; el juego letal
tu alimento, emociones y vibraciones son
  la medicina real.
¡Despierta!
Si el humano, no escucha, no ve, no
  siente, no trasciende
¿para qué existe?
El tiempo irreal asesina
dicen que no se puede olvidar y tampoco
  no buscar
el pasado como no se olvida te aprisiona,
el futuro como es incierto te presiona,
de todos los tiempos solo el presente
  permite vivir libre.
¡Despierta!
Si el humano, no escucha, no ve, no
  siente, no trasciende
¿para qué existe?

Individualism is weakness
harmonious existence needs duality
complimentary opposites give balance
  to life
the union of these extremes sends out
  vibrations
Individualism survives, duality co-exists.
Wake up!
If humans don't listen, don't see, don't
  feel, don't transcend
why exist?
The medicine is you
all illness is created
used as a weapon by transnationals
they poison you – they cure you; a
  lethal game
your food, emotions and vibrations are
  the real medicine.
Wake up!
If humans don't listen, don't see, don't
  feel, don't transcend
why exist?
Unreal time kills
they say that one can't forget nor search
the past that is not forgotten imprisons
  you,
the future with its uncertainty
  pressures you,
of all the times only the present allows
  you to live free.
Wake up!
If humans don't listen, don't see, don't
  feel, don't transcend
why exist?

Land Acknowledgment

Mandie Torres

As Patrick Wolfe (2006) argues, settler colonialism is not just an event of the past; it is deeply embedded in the systems and structures we currently live in and in which we participate. As such, it is important to acknowledge the original caretakers of the lands on which we occupy as a first step to decolonize our world and not perpetuate the ongoing violence against Indigenous peoples. The purpose of a land acknowledgment is to pay respect and honor those who have long looked after the lands which many of us occupy, and not contribute to Indigenous erasure as many of these nations continue to care for these lands. For centuries, many Indigenous groups throughout the world have been actively caring for Mother Earth, and fighting to protect her against destructive colonial and capitalist forces. They have built deep relationships based on reciprocity with the nonhuman world and have engaged in sustainable practices for millennia, from which we can continue to learn and honor.

For example, the Sarayaku community from the Amazon in Ecuador use *Kawsak Sacha* as a concept to live by. *Kawsak Sacha* translates to "the living or breathing jungle/forest," and essentially is the belief that everything in the forest, from the smallest creature to the biggest tree or mountain, are living things with emotions and feelings that we must respect and honor (https://kawsaksacha.org).

During the #NoDAPL Movement that began in 2016, the Lakota saying "Mni Wiconi" or "Water is Life" was heavily circulated by Indigenous water protectors in an effort to bring awareness to halt the construction of a pipeline that runs through the Standing Rock Sioux Indian Reservation, breaking a treaty, potentially poisoning their only water source and the water source of millions of other Americans. Mni Wiconi was used to understand that water itself is literally life, for without it we would not be alive (Privott,

2019). But it also means water has life, and is its own sacred being that the Earth gifts us and therefore we must respect.

Because the purpose of this book is to explore ways of connecting with nature through writing in its many forms, it is especially important that we offer a land acknowledgment before proceeding. Indigenous groups have used different methods of writing, such as song, poetry, storytelling, and much more to speak about their connections with the natural world for millennia, and it is something we as educators can do with our students. As we continuously learn from Indigenous cultures, it is crucial we talk about our connections with the nonhuman world and acknowledge and honor the folks who have been at the forefront of the struggle, resisting and actively fighting against the forces that have been harming Mother Earth for centuries such as settler colonialism, capitalism, white supremacy, and more.

Since many of the contributors to this book currently occupy the land of Los Angeles, we want to recognize and acknowledge "the Gabrielino/Tongva peoples as the traditional land caretakers of Tovaangar (the Los Angeles basin and So. Channel Islands) ... we pay our respects to the Honuukvetam (Ancestors), 'Ahiihirom (Elders), and 'Eyoohiinkem (our relatives/relations) past, present, and emerging" (https://ucla.app.box.com/s/o1texo5yt7qelrelh nzy150jhjugo5mi).

We highly encourage and advocate for all educators, no matter what subject you teach, not only to offer land acknowledgments, as this is merely the first step, but also to engage in further conversations and *actions* that your class can take part in to support Indigenous peoples and to take care of Mother Earth, especially during the climate crisis we currently face.

If you would like to find out who are the traditional caretakers of the land you occupy, visit: https://native-land.ca.

If you would like to learn more about land acknowledgments, their importance, and how to conduct one, we recommend visiting:

- Smithsonian National Museum of the American Indian: https://americanindian.si.edu/nk360/informational/land-acknowl edgment
- Palomar College American Indian Studies Department and Southern California Tribal Chairman's Association's Land Acknowledgment Tool Kit: https://www.csusm.edu/cicsc/land.pdf
- University of California, Los Angeles (UCLA)'s Land Acknowledgment Guidelines: https://ucla.app.box.com/s/o1texo5yt7qelrelhnzy150jh jugo5mi

- The Native Governance Center's Guide to Indigenous Land Acknowledgments: https://nativegov.org/a-guide-to-indigenous-land-acknowledgment/
- *Beyond Land Acknowledgement in Settler Institutions* by Theresa Stewart-Ambo & K. Wayne Yang (2021) https://read.dukeupress.edu/social-text/article-abstract/39/1%20(146)/21/173031/Beyond-Land-Acknowledgment-in-Settler-Institutions

References

Privott, M. (2019). An ethos of responsibility and indigenous women water protectors in the# NoDAPL movement. *American Indian Quarterly*, *43*(1), 74–100.

Wolfe, P. (2006). Settler colonialism and the elimination of the Native. *Journal of Genocide Research*, *8*(4), 387–409.

Acknowledgments

JEFF SHARE

I want to express my gratitude to many people who helped make this book possible. First, I want to thank my students, without whom none of this would have come to fruition. The idea for the book began with them as we learned about ecowriting together in our environmental justice class. My students took the spark and opened my eyes to the powerful pedagogical potential it holds. The ecowriting assignment was initially inspired by Gavin Lamb and his wonderful work that convinced me about the importance of ecowriting. I am honored to have his essay included in this book.

One group of eight students is especially dear to me for their collaborative work peer teaching and writing together every week for months as the pandemic spread across the globe. I am grateful to them for the kindness, passion, generosity, and love they have shared with each other and me. They are: Alejandro Ojeda, Elmer Ortega, Jenifer Ramos, Vanessa Romero, Neida Sandoval-Lopez, Benjamin Thompson, Jazmine Vega Lopez, and María Verónica Valeriano. Supporting this group and me throughout this process has been my teaching assistant, Andrea Gambino, who has infused this work with her commitment and passion to social and environmental justice.

I am also grateful to Justin C. M. Brown for his fabulous cover image that conveys the joy and sense of giving that ecowriting encompasses. The lesson plans are a powerful practical addition to this collection of essays that support teachers with the scaffolding to bring these lessons into the classroom. I am indebted to Andrea Gambino and Sydney Richmond for creating these wonderful lessons with the supporting resources. My two assistants throughout the process are my former student, Mandie Torres, who organized the initial work and also wrote the land acknowledgment, and Andrea Gambino, who brought the work together into one cohesive volume. I want to thank my colleagues at the University of California, Los

Angeles (UCLA) who have passionately supported the environmental justice work we are developing in the School of Education and Information Studies: Dean Christina Christie, Chair Cecilia Rios-Aguilar, Executive Director of Center X Annamarie Francois, Director of the Teacher Education Program Emma Hipolito, Arif Amlani, Director of Program Development, Academic Director of Undergraduate Programs Mitsue Yokota, and Vice Chair of Undergraduate Education Rashmita Mistry. At Peter Lang, I am indebted to Shirley Steinberg for never giving up on this project.

I want to thank the awesome editing skills of my mother, Armony Share, my wife, Laura Vargas, and my former UCLA teaching partner, Sheila Lane. Those three women read the essays and provided essential direction and correction of the texts.

All the authors who have donated their time, energy, and passion to write the essays, poems, and lessons for this book are forever in my debt. It is such a privilege and joy to include writing from so many people I deeply respect from locations across the globe. It has been a great pleasure and honor to work with all the extraordinary people who contributed to this book. Thank you immensely.

Introduction

Jeff Share

Introduction Figure: *This photograph, known as "Earthrise," was taken in 1968 by astronaut William Anders on Apollo 8. Photo Credit: William Anders/NASA via AP*

After 18 days of a space mission, I was convinced that all visible space – the black emptiness, the white, unblinking stars and planets – was lifeless. The thought that life and humankind might be unique in the endless universe depressed me and brought melancholy upon me, and yet, at the same time, compelled me to evaluate everything differently. Nature has been limitlessly kind to us, having helped humankind appear, stand up, and grow stronger. She has generously given us everything she has amassed over the billions of years of inanimate development. We have grown strong and powerful, yet how have we answered this goodness? (Yuri Glazkov, U.S.S.R., p. 84, as cited in Kelley, 1988)

As I looked down, I saw a large river meandering slowly along for miles, passing from one country to another without stopping. I also saw huge forests, extending across several borders. And I watched the extent of one ocean touch the shores of separate continents. Two words leaped to mind as I looked down on all this: commonality and interdependence. We are one world. (John-David Bartoe, U.S.A., p. 86, as cited in Kelley, 1988)

Looking back at Earth from space has inspired deep revelations about the majesty and fragility of this blue pearl floating in a sea of darkness. In the coffee-table book, "The Home Planet," editor Kevin Kelley (1988) shares a collection of spectacular photographs of planet Earth taken from space and accompanied by personal reflections from astronauts and cosmonauts. Kelley states that they report "changes that are powerful and life-transforming, whole-life changes they attribute to the simple experience of looking back at our home planet from the remoteness of space" (p. 1). Frank White calls this the "overview effect," in which the experience of seeing Earth from space can cause "a cognitive and emotional shift in a person's awareness, their consciousness and identity" (Rivera, 2022). The photograph of Earth rising over the barren horizon of the moon, taken from Apollo 8, is one of those images that has changed our perceptions of the world. It is in this spirit of exploration, reflection, and consciousness raising that this book aims to guide us along a journey of discovery to see our vulnerable and finite home anew, inspired to contemplate and deepen our relationships with the natural world.

Teaching about the climate crisis and the devastating impact of environmental abuse, contamination, and extinction is important, yet can be overwhelming and depressing for even the most optimistic. For this reason, I begin my classes by focusing on our positive experiences with nature and the successful work of environmentalists, to develop and celebrate biophilia (the love of nature). We protect what we love. Therefore, cultivating a love of nature is essential for empowering students to become environmental stewards.

Educational Philosopher Nel Noddings (2016) writes about the importance of developing a love of nature that is interdisciplinary, bringing meaning and connections to an array of subject matter and topics that have been artificially separated. To challenge these disconnections, students can benefit from exploring the relationships that unite us. Recognizing our connections to each other, our interdependence with the land and all aspects of the natural world are the basic elements of biophilia. Ecolinguist Gavin Lamb reminds us that our love of nature should not allow for the destruction of one place or people for the benefit of another, as occurs when countries and corporations extract resources and exploit labor from places chosen to be sacrificed, and then dump their waste in those sacrificed zones. Noddings writes, "Love of place should lead us to a love of peace" (p. 19). This involves a local and global commitment to nature and justice that moves beyond profit, nationalism, and borders to embrace our interconnectedness to all ecosystems and people.

Our connections and disconnections to nature is a theme that runs through all the essays in this book, from the opening poem *No Más Humanos [No More Humans]* by Sarawi Andrango to the final student examples of ecowriting. Making this concept even more vital is an understanding of the disconnections that have disrupted our oneness with nature and led to so many of the environmental problems we now face. Bestselling author Naomi Klein (2019) retakes Carolyn Merchant's (1980) research from her book *The Death of Nature* to remind us that "up until the 1600s, the earth was seen as alive, usually taking the form of a mother. Europeans, like indigenous peoples the world over, believed the planet to be a living organism, full of life-giving powers" (p. 59). It is no accident that this idea, which the Inca people still refer to as *Pachamama* (Mother Earth), has disappeared from most Western societies. For centuries, the influences of colonialism, capitalism, and Christendom have separated people from the natural world with the belief that nature is merely a resource to be appropriated and consumed by humans, along with a definition of "progress," based on the extraction and exploitation of people and natural resources.

In her essay, Melissa Greene-Blye describes how the imposition of Western practices, ideologies, and systems has caused great harm to Indigenous peoples and the Earth. The process of colonization that began in the fifteenth century involved classifying, categorizing, and naming places and people that were new to the colonizers. It was a dehumanizing process that turned people and things into objects and commodities that could be controlled and exploited for the benefit of the European colonizers. This process was sanctioned by the Catholic Church through Papal Bulls issued from the Vatican, establishing what is still today legal precedent known as the *Doctrine of Discovery* (Miller et al., 2012; Newcomb, 2008). Colonialism not only separated people from nature but also created the idea of separate human races, with the belief that Christian Europeans were superior to all others. These disconnections from nature and each other are, in many ways, the dawn of the global expansion of white supremacy and the environmental crises we are now facing. Lamb warns about the danger of Amitav Ghosh's *Great Derangement* from nature and the need to write about the natural world in the active voice to guide us toward a *Great Recognition* of our interdependence with nature.

While numerous cultures were impacted and changed by colonization, not all societies lost their connections and beliefs in the importance of living in balance with the natural world, and not all cultures adapted the Western ideals of human domination over nature. Many Indigenous peoples around the world have held on to their relational beliefs and customs as they struggle to preserve their traditional ecological knowledge and sustainable practices that have evolved over millennia. These Indigenous concepts are distinct from an *ego-logical* worldview that emphasizes a hierarchy of power and gender binary in which men are above women, humans above animals, animals above plants, and plants above minerals, ideas represented in the Christian *Great Chain of Being* (Martusewicz et al., 2015). These worldviews conflict, as an *eco-logical* worldview prioritizes the interrelationships of all life without hierarchies that place one species above another. The ecological perspective recognizes all life as systems of reciprocity and relationships of interdependence.

Ecological worldviews can be seen in the language and words used to represent the beliefs and assumptions that communities value. In *Braiding Sweetgrass*, Robin Wall Kimmerer (2013) writes, "When a language dies, so much more than words are lost. Language is the dwelling place of ideas that do not exist anywhere else. It is a prism through which to see the world" (p. 258). The power of language and words is immense, and as students use them to write about their relationships with nature, they gain access to these tools that can transform themselves and the world around them. Lamb encourages us to build a toolkit of environmental keywords to help us understand and tell new stories about the complexity of our relationship with nature. Greene-Blye advocates for sharing stories written by Indigenous authors to explore various ways of seeing and understanding our relationships and responsibilities to the land. We can see this power especially clear when we consider words embedded in Indigenous languages that do not exist in colonial languages, like English or Spanish. The following are two examples of words that express an Indigenous ecological worldview.

Sumak kawsay are words from Quechua, an Indigenous language in the Andes, that are included in Ecuador's 2008 national constitution. As the first country to incorporate the Rights of Nature in their constitution, Ecuadorians chose the words *sumak kawsay* to express the Andean relational worldview that challenges capitalistic notions of development and promotes recognition of intrinsic values of the natural world (Rodríguez Morales, 2022). Following the preamble of Ecuador's constitution, it states: "Decidimos construer ... Una nueva forma de convivencia ciudadana, en diversidad y armonía con la naturaleza, para alcanzar el buen vivir, el *sumak kawsay*;" [English Translation: We hereby decide to build ... A new form of public coexistence, in diversity and in harmony with nature, to achieve the good way of living, the *sumak kawsay*].

> The Spirit of *Aloha* is multi-dimensional. It is an action word that takes many forms. There is a Hawaiian proverb which says *'O ke aloha 'aina ke kuleana i kakou a pau* [Love for the land is our privilege and responsibility]. Hawaiian epistemology has a deep appreciation for richness and fullness of life beyond linear thinking. Hawaiians have a deep respect for all life forms, from the heavens, to the earth, to the land, to the sea. All of these things are our ancestors for whom we have deep *aloha* for. All living things (seen and unseen) have a spirit and consciousness. When we say "*Aloha 'Aina*", we are acknowledging our connection and respect to the land and its people. (Kumu Lehua Hawkins, *The Spirit of Aloha*, personal communication, September 21, 2020)

The idea for creating this book came about from teaching an environmental justice class to undergraduate students at the University of California, Los Angeles (UCLA). For that class, we read an essay by Gavin Lamb about ecowriting to provide a structure and purpose for exploring our relationships with nature. The activity was highly motivating, encouraging students to produce an array of writing, from short stories to poems to a movie script. In part three of this book, we include writing samples created in that class. More important than the final product is the learning that occurs through the *process* of reflecting and creating texts about our relationships with the natural world.

Author Rebecca Solnit asserts the need for us all to become critical listeners and speakers, and conscientious readers and writers, because "stories can give power, or they can take it away." In the twenty-first century, our stories are more likely to be shared through social media, movies, podcasts, video games, and digital networks than the traditional printed book. While this does not mean we should ignore book literacy, it does speak volumes to the need for expanding our understanding of reading and writing to include multiple mediums of communication, such as cartoons, video essays, visual journals, photography, drawings, as well as letters on a page.

In 2021, we celebrated one hundred years since the birth of Brazilian educator and philosopher Paulo Freire, who wrote extensively about the power of literacy to oppress or liberate. He challenged traditional methods of teaching that emphasize rote memorization, something he equated to making deposits of information in a memory bank, a pedagogy better suited to teaching conformity and obedience rather than critical thinking and empowerment. Instead of this "banking" method of teaching, Freire (2010) called for *problem-posing pedagogy* that engages students with problems that they address in the process of learning literacy skills, thereby developing and applying the tools for transformation through engaging with the issues most relevant to them. In their writing about literacy, Paulo Freire and Donaldo Macedo (1987) assert: "Reading the world always precedes reading the word, and reading the word implies continually reading the world" (p. 35). They frame literacy as a process that involves reading as well as writing and rewriting the word and world, as a means to transform it.

For the Love of Nature follows in Freire's footsteps as we embrace the greatest crisis facing humanity today, the one factor that affects everyone on this planet, but not equally. The climate crisis is our existential threat that requires us all to address the current problems and solutions. As educators, we have a significant role to play in educating and empowering students of every age. Education is and has always been political; it has long been used as a tool to control societies or to liberate individuals, to stifle growth or to expand awareness. Students must recognize the seriousness of our environmental crisis and the potential we have to tell new stories and imagine sustainable alternatives. Writing, in all its forms, can be a powerful tool to promote ecological worldviews and unite us in the struggle for environmental justice.

For many years environmental education in the United States lacked an analysis of social justice, a mistake that at times positioned environmentalists against human rights activists. We can overcome this error of the past if we challenge two dominant fallacies: *the myth of universal vulnerability* and *the myth of universal responsibility* (Dunaway, 2015). While all people are impacted by the climate crisis, we are not affected equally. Environmental justice is social justice because of the inequality of environmental impact that has always harmed poor and marginalized communities the worst. Populations with greater resources are temporarily able to escape or lessen many of the effects of rising sea levels, acidification of the oceans, extreme storms, floods, droughts, and fires. The second myth to confront is the mistaken idea that we are all equally to blame. While everyone should do all they can to mitigate their carbon footprint and live sustainable lifestyles, certain corporations and governments have caused much more damage than any individual. The businesses and countries that have benefited from burning fossil fuels and dumping their wastes in sacrifice zones should be held responsible for the externalities of their actions. The fossil fuel companies and industrialized nations that are causing the most damage to the climate have the greatest power and responsibility to significantly reduce the global emissions of greenhouse gases.

We begin this book with a poem from Kichwa poet Sarawi Andrango that she shared with our class as she zoomed in from Ecuador during the Covid lockdown. This is followed by an explanation of land acknowledgments by Mandie Torres, a Chicanx and Central American Studies graduate from UCLA. She highlights the significance of starting with the land and gratitude to those who have taken care of it and continue to live on the land we occupy.

Moving forward, the book is divided into four parts. The first section explores ecowriting and provides a rationale for why it should be integrated into all curricula. Beginning this section, Rebecca Solnit emphasizes the need for new stories that can spark the popular imagination and change our relationship to the physical world. She provides critical hope that rejects the austerity narrative in exchange for an empowering vision of our interconnectedness. Gavin Lamb articulates the power of ecowriting and guides us through strategies for creating ecowriting in which everyone can engage. Lamb takes us on a journey of ecowriting that inspires the senses, explores multiple perspectives, and illuminates the magic of writing. In the following essay, Denise Chapman transports us through stories to other places and times, helping us appreciate the power of storytelling. She applies bell hooks' (2015) notion of *homeplace* and the importance of setting, smells, and seeds for connecting us to land and community. Melissa Greene-Blye continues the celebration of storytelling from Indigenous perspectives and discusses the problems of stereotypical representations that simplify and essentialize Native people. Throughout the book we recognize the importance of storytelling about relationships and responsibilities toward nature.

The second section of the book explores various strategies for teaching ecowriting. Cindy Jenson-Elliot shares her years of experience teaching secondary school students and writing children's books to describe numerous projects that have excited students across the United States. She makes learning come alive through engaging students in hands-on projects, from gardening to creating comic books. Following her essay, Antonio López details the way his students have created video essays that offer multimodal opportunities for academic engagement. He explains his pioneering work in *ecomedia literacy* and paves the way for any teacher to guide their students to create ecomedia video essays. Peaches Hash and Theresa Redmond discuss the power of the arts for counteracting the disconnection with nature that many youth are experiencing. They bring the expressive arts into the process of ecowriting through visual journaling to disrupt alphabetic dominance and create opportunities for bonding creatively with nature. In the next essay, eight of my former students write a collaborative essay about their experiences taking what they learned in our environmental justice class and applying it with their peers in a critical media literacy course. They chose the concept of biophilia and environmental justice and used critical media literacy pedagogy to facilitate a discussion and production of digital poetry with their peers. The final piece of writing for this section is a beautiful reflection by Rose White, one of the most transformative educators I have known.

In the third section, we offer examples of ecowriting from undergraduate students who took my environmental justice class at UCLA and wrote poems, short stories, and letters. Their writing and the experiences we had together in that class were the impetus for this book. Through them, I saw firsthand the power of ecowriting and the passion students develop when encouraged to express their fears, joys, and dreams.

The final section of the book includes lesson plans and resources to support teachers who want to bring these ideas into their classrooms. Sydney Richmond and Andrea Gambino provide detailed instructions for educators interested in the structure and scaffolding for teaching ecowriting. We also include a list of resources for people wanting more ideas and examples.

Kimmerer (2013) explains that many Indigenous cultures operate in a gift economy in which the value of a gift is less as a commodity to be kept and more as an offering that builds relationships based on reciprocity. She states, "Reciprocity is a matter of keeping the gift in motion through self-perpetuating cycles of giving and receiving" (p. 165). It is this notion of relationship building and reciprocal sharing that can encourage us to see ecowriting as a gift, to ourselves, to each other, and to Mother Earth.

Hopefully, this book will inspire the love of nature and guide you to bring ecowriting into your teaching. The variety of examples, from mind-mapping an artifact to creating collaborative digital poems, offers educators, at all grade levels and in any subject matter, an atlas of pathways for exploring ecowriting with students. Now, more than ever, it is crucial to integrate environmental justice into every topic and empower students to research, reflect, and respond to the environmental crises of our time. Ecowriting is a process for learning about environmental justice and taking action through the magic of writing and the power of storytelling.

References

Anders, W. (1968). *Earthrise* [Online image]. NASA via Associated Press.
Dunaway, F. (2015). *Seeing green: The use and abuse of American environmental images.* The University of Chicago Press.
Freire, P. (2010). *Pedagogy of the oppressed* (M. B. Ramos, Trans.). The Continuum International Publishing Group.
Freire, P., & Macedo, D. (1987). *Literacy: Reading the word and the world.* Bergin & Garvey.
hooks, b. (2015). *Yearning: Race, gender, and cultural politics* [Kindle]. Routledge.
Kelley, K. W. (Ed.). (1988). *The home planet.* Addison-Wesley.
Kimmerer, R. W. (2013). *Braiding sweetgrass: Indigenous wisdom, scientific knowledge, and the teachings of plants.* Milkweed.
Klein, N. (2019). *On fire: The (burning) case for a green new deal.* Simon & Schuster.
Martusewicz, R. A., Edmundson, J., & Lupinacci, J. (2015). *EcoJustice education: Toward diverse, democratic, and sustainable communities* (2nd ed.). Routledge.
Merchant, C. (1980). *The death of nature: Women, ecology, and the scientific revolution.* HarperCollins.
Miller, R. J., Ruru, J., Behrendt, L., & Lindberg, T. (2012). *Discovering Indigenous lands: The doctrine of discovery in the English colonies.* Oxford University Press.
Newcomb, S. T. (2008). *Pagans in the promised land: Decoding the doctrine of Christian discovery.* Fulcrum Publishing.
Noddings, N. (2016). Loving and protecting Earth, our home. In K. Winograd. (Ed.), *Education in times of environmental crises: Teaching children to be agents of change* (pp. 14–22). Routledge.
Rivera, E. (2022, October 23). *William Shatner experienced profound grief in space. It was the 'overview effect.'* NPR.org. https://www.npr.org/2022/10/23/1130482740/william-shatner-jeff-bezos-space-travel-overview-effect
Rodríguez Morales, V. V. (2022, April 1). Decolonizing environmental politics: *Sumak kawsay* as a possible moral foundation for green policies. *E-International Relations.* https://www.e-ir.info/2022/04/01/decolonizing-environmental-politics-sumak-kawsay-as-a-possible-moral-foundation-for-greenpolicies/

Part I: Exploring Ecowriting

1. 'If You Win the Popular Imagination, You Change the Game': Why We Need New Stories on Climate

Rebecca Solnit
Copyright Guardian News & Media LTD 2023

So much is happening, both wonderful and terrible – and it matters how we tell it. We can't erase the bad news, but to ignore the good is the route to indifference or despair.

Every crisis is in part a storytelling crisis. This is as true of climate chaos as anything else. We are hemmed in by stories that prevent us from seeing, or believing in, or acting on the possibilities for change. Some are habits of mind, and some are industry propaganda. Sometimes, the situation has changed but the stories haven't, and people follow the old versions, like outdated maps, into dead-ends.

We need to leave the age of fossil fuel behind, swiftly and decisively. But what drives our machines won't change until we change what drives our ideas. The visionary organiser Adrienne Maree Brown (2017) wrote not long ago that there is an element of science fiction in climate action: "We are shaping the future we long for and have not yet experienced. I believe that we are in an imagination battle" (p. 17).

In order to do what the climate crisis demands of us, we have to find stories of a livable future, stories of popular power, stories that motivate people to do what it takes to make the world we need. Perhaps we also need to become better critics and listeners, more careful about what we take in and who's telling it, and what we believe and repeat, because stories can give power – or they can take it away.

To change our relationship to the physical world – to end an era of profligate consumption by the few that has consequences for the many – means changing how we think about pretty much everything: wealth, power, joy, time, space, nature, value, what constitutes a good life, what matters, how change itself happens. As the climate journalist Mary Heglar (2021) writes, we are not short on innovation. "We've got loads of ideas for solar panels and microgrids. While we have all of these pieces, we don't have a picture of how they come together to build a new world. For too long, the climate fight has been limited to scientists and policy experts. While we need their skills, we also need so much more. When I survey the field, it's clear that what we desperately need is more artists" (para. 5).

What the climate crisis is, what we can do about it, and what kind of a world we can have is all about what stories we tell and whose stories are heard. Climate change was a story that fell on mostly indifferent ears when it was first discussed in the mainstream more than 30 years ago. Even a dozen years ago, it was supposed to be happening very slowly and in the distant future. There were a lot of references to "our grandchildren's time". It was a problem that was difficult to grasp – this dispersed, incremental, atmospheric, invisible, global problem with many causes and manifestations, whose solutions are also dispersed and manifold. That voices from the climate movement have finally succeeded in making the vast majority understand it, and many care passionately about it, might be the biggest single victory the movement will have. Because once you've won the popular imagination, you've changed the game and its possible outcomes. But this was a long, slow, arduous process, and misconceptions still abound.

A lot of people don't know that we've largely won the battle to make people aware and concerned. The LA Times ran a well-intentioned editorial last year about how most Americans don't care about climate breakdown (Goldberg, 2022). That was true once, but no longer is. A Pew Research poll in 2020 concluded that two-thirds of Americans wanted to see more government action on climate (Tyson & Kennedy, 2020, para. 4), but last summer the scientific journal *Nature Communications* published a study concluding that most Americans *believe* that only a minority (37–43%) support climate action, when in reality a large majority (66–80%) does (Sparkman et al., 2022). That gap between perceived and actual support undermines motivation and confidence. We need better stories – and sometimes better means more up-to-date.

Outright climate denial – the old story that climate change isn't real – has been rendered largely obsolete (outside social media) by climate-driven catastrophes around the globe and good work by climate activists and journalists. But other stories still stop us from seeing clearly. Greenwashing – the schemes created by fossil fuel corporations and others to portray themselves as on the environment's side while they continue their profitable destruction – is rampant. It's harder to recognise a false friend than an honest enemy, and their false solutions, delaying tactics and empty promises can be confusing for non-experts. Fortunately, as the climate movement has diversified, one new organisation, Clean Creatives (n.d.), focuses specifically on pressuring advertising and PR agencies to stop doing the industry's dirty work. Likewise, climate journalists are exposing how fossil fuel money is funding pseudo-environmental opposition to offshore wind turbines (Thomas & Atkin, 2022).

(As the climate activist and oil policy analyst Antonia Juhasz recently told me, the climate movement is now going after every aspect of the fossil fuel industry, including funding by banks and, via the divestment movement, shares held by investors; donations to politicians; insurers; permits for extraction; transport; refinement; emissions, notably through lawsuits concerning their impact; shutting coal-fired power plants; and pushing for a rapid transition to electrification.)

But we still lack stories that give context. For example, I see people excoriate the mining, principally for lithium and cobalt, that will be an inevitable part of building renewables – turbines, batteries, solar panels, electric machinery – apparently oblivious to the far vaster scale and impact of fossil fuel mining. If you're concerned about mining on indigenous land, about local impacts or labour conditions, I give you the biggest mining operations ever undertaken: for oil, gas, and coal, and the hungry machines that must constantly consume them (Paliwal, 2022).

Extracting material that will be burned up creates the incessant cycle of consumption on which the fossil fuel industry has grown fabulously rich. It creates climate chaos as well as destruction and contamination at every stage of the process. Globally, burning fossil fuels kills almost 9 million people annually (Milman, 2021, para. 1), a death toll larger than any recent war. But that death toll is largely invisible for lack of compelling stories about it.

All mining needs to be done with respect for the land and people in the vicinity, but the impact of mining for renewables needs to be weighed against the far more devastating impact of mining for and burning fossil fuel. The race is on to find battery materials that are more commonly available and less impactful than lithium and cobalt (Balch, 2020), and some of the results look promising. Last summer, Massachusetts Institute of Technology announced an aluminium-sulphur battery is in the works, while a US company is developing one that stores electricity using iron – the so-called "iron-air" battery (Chandler, 2022). Efforts to extract battery materials from long-term coal waste in West Virginia are among the many others under way (Griswold,

2022). And the Inflation Reduction Act includes funding to research better battery materials and domestic US sources (Guido et al., 2022).

Other stories of premature defeat are all too common. In the 400,000-strong 2014 climate march in New York City, one section marched behind a huge banner declaring "WE HAVE THE SOLUTIONS" – but many people still believe we do not. We have the solutions we need in solar and wind (Ambrose, 2021); we just need to build them out and make the transition, fast. Looking to wildly ineffectual carbon sequestration (Gayle, 2022) and other undeveloped technologies as a relevant solution is like ignoring the lifeboats at hand in the hope that fancy new ones are coming when the ship is sinking and speed is of the essence.

One story I frequently encounter frames the possibilities in absolutes: if we can't win everything, then we lose everything. There are so many doom-soaked stories out there – about how civilisation, humanity, even life itself, are scheduled to die out. This apocalyptic thinking is due to another narrative failure: the inability to imagine a world different than the one we currently inhabit.

People without much sense of history imagine the world as static. They assume that if the present order is failing, the system is collapsing, and there is no alternative. A historical imagination equips you to understand that change is ceaseless. You only have to look to the past to see such a world, dramatically different half a century ago, stunningly so a century ago. The UK, for example, ran almost entirely on coal power until the 1960s, and if you had said then that it would have to quit coal, many would have imagined this meant an utter collapse of the energy system, not its transformation (National Grid, n.d.). Even in 2008, the organisation Carbon Brief noted, "four-fifths of the UK's electricity came from fossil fuels. Since then, the UK has cleaned up its electricity mix faster than any other major world economy. Coal-fired power has virtually disappeared and even gas use is down by a quarter. Instead, the country now gets more than half (Rose Morley, n.d.) of its electricity from low-carbon sources, such as solar, wind and nuclear" (Evans & Pearce, n.d.). Scotland already generates nearly all the electricity it needs from renewables (Scottish Renewables, n.d.).

While I often hear people casually assert that our world is doomed, no reputable scientist makes such claims. Most are deeply worried, but far from hopeless. There are already profound losses, but our action or inaction determine how much more loss will occur, and whose it will be, and some repair is possible. Efforts sufficient to reduce the amount of carbon dioxide in the atmosphere could lower temperatures and reverse some aspects of climate breakdown (Hertsgaard et al., 2022).

Even the journalist David Wallace-Wells (2022), who rose to fame with a deeply pessimistic book about climate a few years ago, has shifted his view. He currently describes a future somewhere between the best- and worst-case scenarios, a future "with the most terrifying predictions made improbable by decarbonisation and the most hopeful ones practically foreclosed by tragic delay. The window of possible climate futures is narrowing, and as a result, we are getting a clearer sense of what's to come: a new world, full of disruption ... yet mercifully short of true climate apocalypse" (para. 4).

A climate story we urgently need is one that exposes who is actually responsible for climate chaos. It's been popular to say that we are all responsible, but Oxfam reports that over the past 25 years, the carbon impact of the top 1% of the wealthiest human beings was twice that of the bottom 50%, so responsibility for the impact and the capacity to make change is currently distributed very unevenly (Clifford, 2021).

By saying "we are all responsible", we avoid the fact that the global majority of us don't need to change much, but a minority needs to change a lot. This is also a reminder that the idea that we need to renounce our luxuries and live more simply doesn't really apply to the majority of human beings outside what we could perhaps call the overdeveloped world. What is true of Beverly Hills is not true of the majority from Bangladesh to Bolivia.

When it comes to who's harming the climate, it's also been popular to focus on individual contributions. The fossil fuel industry likes the narrative of personal responsibility as a way to keep us scrutinising ourselves and one another, rather than them. They've promoted the concept of climate footprints as a way to keep the focus on us and not them, and it's worked. Usually if I ask people what they're doing about the climate emergency, most will talk about what they're not consuming or doing – but these will never add up to the speed and scale of change needed to change the system.

One of the goals of system change is to supersede individual virtue. Just as you no longer have to opt in to buying a car with seatbelts or ask for the no-smoking section on the train or restaurant, at some point in the near future you won't have to opt into travelling in an electric car or bus, or living or working in all-electric buildings. Electrification will have happened because of the collective action that takes shape as policy and regulation.

Last year, the veteran environmentalist Bill McKibben (2022b) wrote a brilliant analysis pointing out that if you have money in one of the banks funding fossil fuels – especially, in the US, Wells Fargo, Chase, Citi, and Bank of America – your retirement funds or savings account may have a much larger climate footprint than you do (see *Reclaim Finance*, 2022). The impact of your diet and how you get to work may pale in comparison to the impact of your money in the bank. The vegan on the bicycle may still be contributing to climate chaos if her life savings are in a bank lending her money to the fossil fuel industry.

Individual impact, leaving the ultra-wealthy aside, matters mostly in the aggregate. And in aggregate we can change that. On 21 March, McKibben (2022a), via his new climate group Third Act (on whose advisory board I sit), and dozens of other climate groups will be organising actions by people with money in, or credit cards from, the key US banks, to try to force those institutions to stop funding fossil fuels. Our greatest power lies in our roles as citizens, not consumers, when we can band together to collectively change how our world works.

Various campaigns around the world have focused on fossil finance, with significant successes behind them, and much more to achieve ahead. The climate movement has become far more sophisticated and precise in its targets in recent years. It's doing a brilliant job; it just needs enough people and resources behind it to be more powerful than the status quo.

Last year, I took three activists who were formerly part of the Sunrise Movement, a youth group campaigning to address climate breakdown, to see the 1991 film Terminator 2 at a cinema. The film was as great as I remembered, not least because the lead character, Linda Hamilton playing a ferocious young mother, chooses as her motto "no fate but what we make". In that movie, the future has come back to meddle with the present through the sci-fi technologies of time travel and robot-warrior terminators. We see how actions in the present shape the future through tremendous battles over what that future will be. This is, of course, just as true in real life. We don't get terminators and other time-travellers to tell us what the consequences of our actions are, but they still have consequences. You ban the insecticide DDT, and a lot of bird species stop dying out. You ban chlorofluorocarbons, and the hole in the ozone layer stops growing (Milman, 2023).

In another way, Terminator 2 is less useful as a lens for thinking about the climate crisis. It's part of the conventions of storytelling in film – and comics, fiction, graphic novels and too many news narratives – that tells us that the world can only be saved by exceptional individuals, often loners, whose gifts are often the capacity to inflict and endure extreme violence. Linda Hamilton and co-star Arnold Schwarzenegger shoot, clobber, crush, outrun and outfight everything thrown at them, and that's their celebrated skillset, along with a bit of dry humour.

Humour aside, this has little to do with how the world really gets changed most of the time. The skills of real-world superheroes are solidarity, strategy, patience, persistence, vision and the ability to inspire hope in others. The rescuers we need are mostly not individuals, they are collectives – movements, coalitions, campaigns, civil society. Within those groups there may be someone with an exceptional gift for motivating others, but even the world's greatest conductor needs an orchestra. One person cannot do much; a movement can topple a regime. We are sadly lacking stories in which collective actions or the patient determination of organisers is what changes the world.

Another thing we get from our films and fictions is the expectation of a single solution and a clear resolution to our problems: a sudden victory, a celebration, and the trouble is over. The climate crisis does not fit easily into this format. Ceasing to extract and burn fossil fuel is central, but there is no single solution. Protecting carbon-sequestering peat bogs, forests and grasslands also matters; so does transforming high-impact materials such as cement (Watts, 2019); implementing better design for buildings, transport and cities, and addressing soil conservation, farming and food production and consumption. There are milestones and important goals, but the familiar Hollywood ending – crossing the finish line to wrap up the story – doesn't describe this reality.

Change often functions more like a relay race, with new protagonists picking up where the last left off. In 2019, a Berkeley city councilwoman decided to propose banning fossil-gas connections in new construction, and it was passed by the council unanimously (Cagle, 2019). This small city's commitment to all-electric new buildings could seem insignificant, but more than 50 other California municipalities picked it up, as did the city of New York. The state of New York failed to pass a similar measure, but Washington state succeeded, and the idea that new construction should not include gas has spread internationally.

Such relay races have long been how human rights campaigns work: a good protest, campaign, or even piece of legislation can introduce new ideas that do their own work in the world at large. Even failed campaigns may succeed in opening the path for later change. The Green New Deal did not pass in the US Senate, but it became a template for the Biden administration's climate legislation, and shifted the conversation about what is possible. It led the way to the Inflation Reduction Act (Lowenstein, 2022), the biggest climate bill the United States has ever passed. Opponents of environmental action often say it is killing jobs; the Green New Deal did a lot to change that story by portraying climate action as a job creator.

Recognising the reality of climate breakdown means recognising the interconnectedness of all things. That connection brings obligation: to respect nature, to build domestic regulation and international treaties that protect what's needed, to limit the freedom of the individual in the name of the wellbeing of the collective. This is, of course, a worldview in direct contrast with free-market fundamentalism and libertarianism. Even the facts of climate science are ideologically offensive to people committed to individual freedom without accountability, let alone the demands created by treaties and regulations.

Responsibility and obligation are dismal words in mainstream culture, so perhaps there will be other stories that recognise this process as reciprocity and relationship, in which we give back, in gratitude and respect for all the Earth does for us. Even short of that, we can recognise our self-interest in maintaining the system on which life depends.

If news is the daily report on what's just happened, we need a way of pulling back from individual events, to see the broad context of how it happened. If you only tell short-term stories, it all becomes kind of meaningless. Martin Luther King Jr. (1967) said: "The arc of the moral universe is long, but it bends toward justice" (p. 62). We've seen it bend a lot of ways in recent years, toward and away from justice, but it takes time just to see it bend at all. You need benchmarks or memories of how things used to be even to see change of any kind, including climate change.

The South Pacific climate activist and poet Julian Aguon recently declared that Indigenous peoples "have a unique capacity to resist despair through connection to collective memory, and just might be our best hope to build a new world rooted in reciprocity and mutual respect – for the Earth and for each other". That emphasis on collective memory suggests that a strong sense of the past allows for a strong sense of the future, that remembering difficulty and transformation equips us to face them again.

One of the things that buoys me up is the long arc of change in renewable technology. Mostly what you see in the news about renewables is short-term: stories on the latest drop in price, or proliferation of solar and wind over the past year or two. If you enlarge your time frame, you see that those annual changes have amounted to an astonishing plummet in prices and rise in efficiency and global use, compounded by innovations in materials and storage.

Twenty years ago we did not have constructive ways to leave the age of fossil fuel behind. Now we do. And the solutions keep getting better. In 2021, the organisation Carbon Tracker put out a report that showed current technology could produce 100 times as much electricity from solar and wind than current global demand. The report concludes: "The technical and economic barriers have been crossed and the only impediment to change is political." At the end of the last millennium, those barriers seemed insurmountable.

The change is revolutionary, but the revolution was too slow to be visible to most.

The report continues: "At the current 15–20% growth rates of solar and wind, fossil fuels will be pushed out of the electricity sector by the mid-2030s and out of total energy supply by 2050. The unlocking of energy reserves 100 times our current demand creates new possibilities for cheaper energy and more local jobs in a more equitable world, with far less environmental stress" (para. 7).

We tend to think utopias are unbelievable, but this is a sober-minded thinktank focused on climate and energy politics. The report made little impact on the general public. Because the energy revolution has been incremental, there's been no single breakthrough moment. Yet it adds up to an encouraging, and even astonishing narrative.

On the other hand, people find grim narratives all too believable, whether or not they are grounded in fact. We are still inundated by harmful, as well as untrue, stories about climate and the future. Prophecies can be self-fulfilling: if you insist that we cannot possibly win, you pit yourself against the possibility of victory and the people trying to achieve it.

There's yet another narrative that's persisted at least since the invention of compact fluorescent lightbulbs and the Toyota Prius: that we must renounce abundance and enter an age of austerity. It's all in the telling. To consider our age an age of abundance, you have to be counting sheer accumulated stuff and ignoring how it is distributed. That is, we live in an age of extreme wealth for some, and desperation for the many. But there's another way to count wealth and abundance – as hope for the future, safety and public confidence, emotional wellbeing, love and friendship and strong social networks, meaningful work and purposeful lives, equality and justice and inclusion.

Early on, we heard that renewables were very expensive – this was part of the austerity narrative, or an excuse for not making the transition. But improvements in design and economies of scale are among the factors making them the cheapest form of electricity almost everywhere on earth. There's no reason to think the innovations of design and economic improvements are all behind us; I suspect they're mostly ahead of us.

Engineer and energy expert Saul Griffith (2021) recently wrote: "Most people believe a clean-energy future will require everyone to make do with less, but it actually means we can have better things" (p. 3). The old story was that we couldn't afford to do what the climate emergency required. The new one is that it would not only be ecologically devastating, but more expensive not to. Renewables are on the way to being cheaper than fossil fuel; in many places, they already are (Roser, 2020). Texas and Iowa get a huge amount of their electricity from wind because it makes economic sense, not because these red states are passionate about addressing the climate crisis. Over their lifetime, electric cars work out to be cheaper than internal combustion cars because charging and maintaining them is cheaper. And of course these two examples don't include the indirect effects of burning fossil fuels on human health and the climate.

A lot of people tend to measure climate action in terms of huge national or international news events, but the change that matters is often happening at local and regional and other levels. A university divests; a state sets a date for ending the sale of new petrol cars; a city passes a measure mandating all-electric new buildings; ground is broken on a major solar installation; a state or country sets a new record for percentage of wind power in its energy mix; a pipeline or gas terminal or drilling site gets cancelled; a carbon-sequestering forest or peat bog gets protected status; a coal plant closes (The Flow Country, n.d.).

This does not erase all the bad news, about continuing breakdown of natural systems and its toll on human lives and impact on a livable future, but it does contextualise them as crises we can respond to if we choose to. So much is happening, both wonderful and terrible, and it adds up to more stories than almost anyone can take in. But the overarching frameworks in which we receive them matter, and so do the critical skills to recognise, choose, and change stories.

The climate crisis is a problem with no single solution, but many, just as there is no one saviour, but many protagonists in the struggle. In 2019, Swedish climate activist Greta Thunberg said we must embrace "cathedral thinking", adding: "We must lay the foundation while we may not know exactly how to build the ceiling." The speculative fiction writer Octavia Butler included this passage in one of her essays (Common Good Collective, n.d.):

"OK," *the young man challenged. "So what's the answer?"*

"There isn't one," I told him.

"No answer? You mean we're just doomed?" He smiled as though he thought this might be a joke.

"No," I said. "I mean there's no single answer that will solve all of our future problems. There's no magic bullet. Instead there are thousands of answers – at least. You can be one of them if you choose to be."

References

Ambrose, J. (2021, June 23). Most new wind and solar projects will be cheaper than coal, report finds. *The Guardian*. https://www.theguardian.com/environment/2021/jun/23/most-new-wind-solar-projects-cheaper-than-coal-report

Balch, O. (2020, December 8). The curse of 'white oil': Electric vehicles dirty secret. *The Guardian*. https://www.theguardian.com/news/2020/dec/08/the-curse-of-white-oil-electric-vehicles-dirty-secret-lithium

Brown, A. M. (2017). *Emergent strategy: Shaping change, changing worlds*. AK Press.

Cagle, S. (2019, July 23). Berkeley became first US city to ban natural gas. Here's what that may mean for the future. *The Guardian*. https://www.theguardian.com/environment/2019/jul/23/berkeley-natural-gas-ban-environment

Carbon Tracker. (2021, April 23). *The sky's the limit: Solar and wind energy potential is 100 times as much as global energy demand*. https://carbontracker.org/reports/the-skys-the-limit-solar-wind/?mbid=&utm_source=nl&utm_brand=tny&utm_mailing=TNY_Climate_042821&utm_campaign=aud-dev&utm_medium=email&bxid=5bd673de24c17c104800a1c0&cndid=32390035&hasha=9f3d45e07fc910bc25840bc92486bca2&hashb=b84e32208904f47e6d9dc356bcb2e7e43d610f98&hashc=b3d65a0ad31dee847d668fe14dac8b9ffbcd841a156dc78401d1c3158e6ec796&esrc=&utm_term=TNY_ClimateCrisis

Chandler, D. L. (2022, August 24). *A new concept for low-cost batteries: Made from inexpensive, abundant materials, an aluminum-sulfur battery could provide low-cost backup storage for renewable energy sources*. MIT News Office. https://news.mit.edu/2022/aluminum-sulfur-battery-0824

Clean Creatives. (n.d.). *The future of creativity is green*. https://cleancreatives.org/

Clifford, C. (2021, January 27). The '1%' are the main drivers of climate change, but it hits the poor the hardest: Oxfam report. CNBC. https://www.cnbc.com/2021/01/26/oxfam-report-the-global-wealthy-are-main-drivers-of-climate-change.html

Common Good Collective. (n.d.). *A few rules for predicting the future by Octavia E. Butler*. https://commongood.cc/reader/a-few-rules-for-predicting-the-future-by-octavia-e-butler/

Evans, S., & Pearce, R. (n.d.). *How the UK transformed its electricity supply in just a decade*. Carbon Brief. https://interactive.carbonbrief.org/how-uk-transformed-electricity-supply-decade/

The Flow Country. (n.d.). *Restoring the flows*. https://www.theflowcountry.org.uk/flow-facts/flow-fact-4/#:~:text=There%20was%20a%20major%20push,project%20through%20Peatland%20Action%20funding.

Gayle, D. (2022, September 1). Carbon capture is not a solution to net zero emissions plans, report says. *The Guardian*. https://www.theguardian.com/environment/2022/sep/01/carbon-capture-is-not-a-solution-to-net-zero-emissions-plans-report-says

Goldberg, N. (2022, September 22). Column: Americans don't care about climate change. Here's how to wake them up. *The Los Angeles Times.* https://www.latimes.com/opinion/story/2022-09-22/climate-change-concern-marketing

Griffith, S. (2021). *Electrify: An optimist's playbook for our clean energy future.* MIT Press.

Griswold, E. (2022, August 26). Could coal waste be used to make sustainable batteries? *The New Yorker.* https://www.newyorker.com/news/us-journal/could-coal-waste-be-used-to-make-sustainable-batteries?fbclid=IwAR05qx3tTSd-VZUw5CGKK4VCrgzhwLp2I84LhjVPgoPRn1nAdihChzbUmRZI

Guido, V., Iyer, N., & Lezak, S. (2022, October 12). *How the inflation reduction act will spur a revolution in EV battery supply chains.* ARMI. https://rmi.org/how-the-inflation-reduction-act-will-spur-a-revolution-in-ev-battery-supply-chains/

Hegler, M. A. (2021, October 24). To build a beautiful world, you first have to imagine it: Looking at the climate fight, it's clear what we desperately need is more artists. *The Nation.* https://www.thenation.com/article/environment/climate-world-building/

Hertsgaard, M., Huw, S., & Mann, M. E. (2022, February 23). How a little-discussed revision of climate science could help avert doom. *The Washington Post.* https://www.washingtonpost.com/outlook/2022/02/23/warming-timeline-carbon-budget-climate-science/

King Jr., M. L. (1967). *Where do we go from here: Chaos or community?* Beacon Press.

Klein, N. (2022, October 18). Greenwashing a police state: The truth behind Egypt's COP27 masquerade. *The Guardian.* https://www.theguardian.com/environment/2022/oct/18/greenwashing-police-state-egypt-cop27-masquerade-naomi-klein-climate-crisis

Lowenstein, A. (2022, November 6). Biden's climate bill victory was hard won. Now, the real battle starts. *The Guardian.* https://www.theguardian.com/global-development/2022/nov/06/inflation-reduction-act-climate-crisis-congress

McKibben, B. (2022a, December). *32123!: Big banks are driving the climate crisis, so we're pushing back.* The Crucial Years. https://billmckibben.substack.com/p/32123

McKibben, B. (2022b, May 20). *Your money is your carbon.* The Crucial Years. https://billmckibben.substack.com/p/your-money-is-your-carbon

Milman, O. (2021, February 9). 'Invisible killer': Fossil fuels caused 8.7m deaths globally in 2018, research finds. *The Guardian.* https://www.theguardian.com/environment/2021/feb/09/fossil-fuels-pollution-deaths-research

Milman, O. (2023, January 9). Earth's ozone layer on course to be healed within decades, UN report finds. *The Guardian.* https://www.theguardian.com/environment/2023/jan/09/ozone-layer-healed-within-decades-un-report

National Grid. (n.d.). *The history of energy in the UK.* National Grid Group. https://www.nationalgrid.com/stories/energy-explained/history-of-energy-UK

Paliwal, A. (2022, December 20). 'It was a set-up, we were fooled': The coal mine that ate an Indian village. *The Guardian.* https://www.theguardian.com/environment/2022/dec/20/india-adani-coal-mine-kete-hasdeo-arand-forest-displaced-villages

Reclaim Finance. (2022, March 30). *New report: World's biggest banks continued to pour billions into fossil fuel expansion in 2021, Press release by Rainforest Action Network, BankTrack, Indigenous Environmental Network, Oil Change International, Reclaim Finance, Sierra Club, and Urgewald.* https://reclaimfinance.org/site/en/2022/03/30/new-report-worlds-biggest-banks-continued-to-pour-billions-into-fossil-fuel-expansion-in-2021/

Rose Morley, K. (n.d.). *National grid: Live.* https://grid.iamkate.com/

Roser, M. (2020, December 1). *Why did renewables become so cheap so fast?* Our World in Data. https://ourworldindata.org/cheap-renewables-growth

Scottish Renewables. (n.d.). *Statistics: Energy consumption by sector.* https://www.scottishrenewables.com/our-industry/statistics

Sparkman, G., Geiger, N., & Weber, E. U. (2022). Americans experience a false social reality by underestimating popular climate policy support by nearly half. *Nature Communications, 13*(4779), 1–9.

Thomas, M., & Atkin, E. (2022, November 29). *The fossil fuel industry's deceptive campaign to kill offshore wind.* Distilled. https://www.distilled.earth/p/the-fossil-fuel-industrys-deceptive

Tyson, A., & Kennedy, B. (2020, June 23). *Two-thirds of Americans think government should do more on climate.* Pew Research Center. https://www.pewresearch.org/science/2020/06/23/two-thirds-of-americans-think-government-should-do-more-on-climate/

Wallace-Wells, D. (2022, October 26). Beyond catastrophe: A new climate reality is coming into view. *New York Times Magazine.* https://www.nytimes.com/interactive/2022/10/26/magazine/climate-change-warming-world.html

Watts, J. (2019, February 25). Concrete: The most destructive material on Earth. *The Guardian.* https://www.theguardian.com/cities/2019/feb/25/concrete-the-most-destructive-material-on-earth

2. Ecowriting: A Fieldguide

GAVIN LAMB

Introduction

What stories do we tell about our relationship to the natural world? And what consequences do these stories have for how well we address the monumental environmental and social challenges of our time?

Ecowriting is a phrase I use to describe a loose collection of old and new approaches to nature writing and 'environmental storytelling' that take up these challenges. They do so by taking up writing as, to borrow the words of Thom van Dooren and Deborah Bird Rose (2016), a "mode of knowing, engaging, and storytelling that recognizes the meaningful lives of others and that, in so doing, enlivens our capacity to respond to them by singing up their character or ethos" (p. 77). In this sense, I envision ecowriting as a mode of storytelling to restore a recognition of interconnectedness among all human and nonhuman beings in a time of accelerating socio-ecological crises around the world. Perhaps, most of all, it's an approach to writing that continually asks whose stories are being excluded or erased altogether, and deliberately pushes back against ecologically and socially destructive stories being told. In my view, ecowriting, as Robert Macfarlane (2019) aptly puts it, is a way of 'counter-mapping': "remapping a landscape with a view to pushing back against power."

Ecowriting is like a compost heap of stories; a tangle of personal, social, cultural, ecological, political, economic, technological, and scientific stories that provides the detritus to become hummus for something new. What might these new ecostories look like? Luckily, a range of writers – from journalists and activists to researchers and novelists from around the world – are telling them. They show us that ecowriting is not just about reframing how we 'talk' about nature. For these writers, we can't continue writing about a nature 'out there' in the 'environment' that isn't also part of us 'in here.' The interwoven crises of climate change and environmental injustice make this beyond clear. So, now it is time to look to storytellers writing from a place of reckoning with what's been lost, or all we might still lose, but above all, 'all we can save' (Johnson & Wilkinson, 2020).

On the one hand, ecowriting writes *against* a deeply entrenched Western intellectual dualism that would separate humans from nature, and instead writes *for* a vision that is responsive to the interdependence of human beings and more-than-human beings. "We are in and of the world," writes cultural ecologist David Abram (2010), "materially embedded in the same rain-drenched field that the rocks and the ravens inhabit . . . All our knowledge, in this sense, is carnal knowledge, born of the encounter between our flesh and the cacophonous landscape we inhabit" (p. 72).

If this is true, shouldn't all the environmental stories we tell be embedded in the landscapes we inhabit? The challenge for ecowriters is that the magic of writing comes from how we use writing as a tool. And like all tools, it can be harnessed for both positive and negative goals: to amplify our sense of interdependence with the assemblage of beings and places we depend on for our well-being, or instead, conceal and erase our connections to all the "shadow places," as Val Plumwood (2008) calls them, those sacrificial places that support our flourishing, yet that dominant stories compel us to ignore.

Unfortunately, the dominant stories being told are still rooted in a modernist ecocultural framework based on mass consumption, human mastery over nature, and infinite growth. Now, more than ever, we need alternative stories to this dominant framework, stories that remind us of our embedding in the landscapes we inhabit, and our interconnectedness with human and nonhuman beings on this planet.

Here, I provide a few ideas on how to get started. As an introduction to ecowriting, this essay is divided into two parts. The first part, "strategies," examines three higher-level strategies to guide your ecowriting practice, calling to mind some helpful principles. The second part, "tactics," digs into a few actionable tips for ecowriting. These are not meant to define or constrain what ecowriting is, but to offer some guiding principles and ethos to begin the journey.

It's in the combination of these strategies and tactics that this fieldguide can be of most use to you in developing your own ecowriting practice.

As you read through the strategies and tactics, you'll hopefully discover new writers you hadn't heard of before, and also some you might be familiar with. For example, you'll find advice from journalists like Michael Pollan (2007) who has inspired many through his storytelling about the agency of plants. But writers like Pollan rarely talk about how they write, so I've pulled much of this advice not only from analyzing the structure that writers like him deploy, but also from bits of wisdom they've shared about their writing process in interviews.

In addition, I've also drawn on approaches to environmental storytelling from science fiction writers, novelists, anthropologists, geologists, historians, conservationists, psychologists, and more. My hope is that you discover particular writers who resonate with you and inspire your own thinking about environmental storytelling.

Here, I draw inspiration from artist and writer Deborah J. Haynes' (2003) suggestion to find a philosopher to write with. "The reason for writing is simple," Haynes says. "Through writing you will learn what you think and you will come to know yourself. Write to find out what you think. Unless you know what you think, you will always be subject to the will of others." In her unconventional self-help book for creatives, *Art Lessons: Meditations on the Creative Life*, Haynes recommends choosing one philosopher to write with. It doesn't really matter which philosopher, how many, or even what counts as a philosopher to you. The point of this exercise, states Haynes, is that simply choosing someone whose ideas you can engage with can help you start a useful dialogue in your mind.

So, you can think of this fieldguide as a collection of ecophilosophers you can begin a personal dialogue with. And like a garden of ideas, I invite you to take a walk through their ideas about environmental storytelling and observe, tend to, and pick any that catch your eye, helping you develop your own distinct ecowriting practice.

With all this said, what's the point of ecowriting? Can good environmental storytelling actually get individuals, groups, organizations, governments, and maybe even society, to act on the monumental ecosocial crises we face in this moment? Can it compel the kind of radical systemic change we need urgently? Do the words we use even have the power to inspire such change? I think they do. And if that's true, then it's important to find the strategies we can use as more effective tools to expand knowledge, shift attitudes and promote positive change in the world. I hope you find some of these strategies and tactics useful to inspire your own work as an ecowriter, and ultimately, to contribute to a more just, joyful, and life-sustaining future.

Strategies

Connecting the Dots

How can we tell stories that connect the dots from our daily lives to the wider networks of people and places we become connected to through our everyday choices and actions?

When it comes to the food we eat, the writer Michael Pollan (2007) has made a name for himself doing just this: connecting the dots of the food we eat by tracing their histories from our plate to the farm, and finally to the obfuscated agricultural industries that allow for our daily meals. While giving writing advice is not something he does – editing, reediting, and editing it all over again is his main advice for aspiring writers – a phrase, or ethos, that has inspired his writing for years comes from the environmental philosopher/theologian Wendell Berry (2010): "To eat is an agricultural act."

Pollan writes, "When Berry says 'eating is an agricultural act,' that's a very empowering statement. He's saying you have political power in your everyday actions. When you decide what you're going to eat, what you're going to buy, you have real influence. That's why this idea has the potential to resonate with so many people. It's certainly one of the reasons it's resonated with me: I know I can act today. Three times" (Fassler, 2013).

But to connect the dots, we need tools to do so.

One such set of tools comes from the emerging field of ecolinguistics. Salience patterns are linguistic patterns for representing the world that allow us to push back against erasure patterns. Erasure patterns, by contrast, are linguistic patterns that work to obscure the links between the dots we wish to connect (Stibbe, 2015).

In other words, the power of Berry's (2010) statement, "eating is an agricultural act," comes from how it renders visible to us – makes salient – how every bite we take links us up to a vast network of industrial agriculture (at least for many of us living in modern society). This is a sprawling food network many of us don't know much about. Making matters worse, according to Pollan (2007), is that there is a "vast conspiracy of silence" that whirls around much of the food we buy at the supermarket: maybe it's the information about where an animal came from or who grew the vegetables, but much of this history is strategically made invisible to us.

Or, maybe what we are allowed to see is greenwashed through the filter of "supermarket pastoral" advertising, encouraging us to think the happy cow grazing on the pasture on a milk carton or cheese package is reality.

"All this makes it difficult to act on Berry's injunction," Pollan writes. "In fact, capitalism depends on erecting these screens, strives to defeat efforts to see the lines of connection between you, and the farmworker who picked your strawberries, and the corporation that delivered them to your door" (Fassler, 2013).

An important insight Wendell Berry gives us is that we inhabit an economic system that more often than not tries to keep us from connecting the dots – hiding from us that whatever we buy implicates us in a chain of interrelationships with other people, animals, places, and ecosystems. That's a challenging idea to be walking around with all day, Pollan suggests. We tend to avoid thinking about how someone on the other side of the planet may have been exploited to make the products we enjoy, from iPhones and laptops to our food and clothes. Even so, Berry's adage calls on us to start thinking about these dots, and how we might connect them.

'Connecting the dots' is a way of writing that tells us: since eating is an agricultural act, eating is a political act, too. Connecting the dots isn't always easy, though. For Pollan, one strategy has been to expand his understanding of the webs of commerce he is suspended in by interviewing other people caught up in these webs and researching their history in an attempt "to trace the whole long chain," from his individual actions to the wider world.

"In a way, all my writing about food has been about connecting dots in the way Berry asks of us," says Pollan.

> It's why, when I write about something like the meat industry, I try to trace the whole long chain: from your plate to the feedlot, and from there to the corn field, and from there to the oil fields in the Middle East. Berry reminds us that we're part of a food system, and we need to think about our eating with this fact – and its implications – in mind ... We stand to gain so much by connecting these dots. (Fassler, 2013)

As an ecowriting tool for environmental storytellers, connecting the dots involves tacking back and forth among three questions: (1) What is the concrete action I'm interested in telling a story about? (e.g., growing a garden, flying to Hawai'i, buying a cup of coffee), (2) What are the other actions, beings, events and places this action is connected to that make it more or less possible?, and (3) How do different ways of telling a story about this action make the many worldly connections it depends on either more salient or more obscure?

I find that starting with very particular, concrete action and tracing outward from there in space and time is a helpful way to tell stories of interconnectedness, stories that reveal the nexus of life that make our everyday actions possible or impossible. Usually, for me, I need to tell more than one story about the same action, and from different perspectives, to reveal this wider nexus of histories and activities my everyday actions are entangled with.

Nature in the Active Voice

Have you ever had the experience of being next to or touching something you thought was inanimate but turned out to be alive? When a motionless branch turns out to be a snake? Or a floating log an alligator? I've witnessed this when tourists walking along the beach in Hawai'i stroll right past what seems to be a larger round rock, only to discover moments later the rock is actually alive: startled, they shout, 'sea turtle!' in gawking awe.

Amitav Ghosh (2018), the environmental novelist, writes in his wonderful book *The Great Derangement: Climate Change and the Unthinkable*: "Who can forget those moments when something that seems inanimate turns out to be vitally, even dangerously alive?" (p. 3).

Ghosh goes on to write that the filmmakers of *The Empire Strikes Back* must have been imagining a similar moment in their experience when they made the memorable 'this is no cave' scene.

To evade an attack, Han Solo lands the Millennium Falcon inside an asteroid cave. But a few moments after landing, the ground begins to quake. Fearing the cave will crumble, they make their escape.

"The cave is collapsing!" Princess Leia exclaims as the toothy mouth of the 'cave' begins to shut. "This is no cave" Han Solo replies, as they barely scrape through the shrinking opening. It soon becomes clear: they're not escaping from any ordinary cave but from the gut of a sleeping space monster!

For Amitav Ghosh, this Star Wars scene reveals a deeper meaning about the place of human beings in a more-than-human world:

> The humans of the future will surely understand, knowing what they presumably will know about the history of their forebears on Earth, that only in one, very brief era, lasting less than three centuries, did a significant number of their kind believe that planets and asteroids are inert.

People of the future will regard past human beings with astonishment. "There was a strange time," a future Han Solo might say to a future Leia, "when people actually believed living planets were just inanimate rocks."

And then, the climate crisis inflicted a sudden shock of recognition, interrupting this strange period of time people of the future will remember as *The Great Derangement* – that strange string of centuries when human beings (or, rather, an influential subset of human beings) refused to recognize they were walking on a living planet, like the cave below Han Solo's feet when it occurred to him the ground was actually alive, dangerously alive.

As Amitav Ghosh writes, "Quite possibly, then, this era, which so congratulates itself on its self-awareness, will come to be known as the time of the Great Derangement." However, a growing number of writers are telling stories of nature in the active voice to counteract this dominant model of nature-as-backdrop, as if the planet were simply an inert stage for human dramas to unfold.

Ghosh argues that environmental storytelling, whether in fiction or non-fiction, has an important role to play in helping us to tell stories of nature in the active voice, to avert the worst of the climate crisis, since stories aren't just about the world, but have consequences for how we act in the world.

Will the future Leias and Han Solos of the world call this strange period in history, not the Great Derangement, but the Great Recognition?

Can better environmental storytelling help us to usher in, even if only a little bit faster, a moment in history when we finally recognize our interdependence with a more-than-human world: when we finally remember that "collaborative survival requires cross-species coordination," as the environmental anthropologist Anna Tsing (2015) puts it?

I think so.

By writing nature in the active voice, I believe environmental storytellers still have much to contribute to guiding us toward Ghosh's "Great Recognition:" when we stop acting like the planet is just a container for human action. Luckily, there are many ecowriters who are already showing us alternatives, writers like the ecofeminist philosopher Val Plumwood (2010):

> Writers are amongst the foremost of those who can help us to think differently. Of course, artistic integrity, honesty and truthfulness to experience are crucial in any re-discovery of 'tongues in trees.' I am not talking about inventing fairies at the bottom of the garden. It's a matter of being open to experiences of nature as powerful, agentic and creative, making space in our culture for an animating sensibility and vocabulary.

Reminding

> "We can be ethical only in relation to something we can see, feel, understand, love, or otherwise have faith in."
> – Aldo Leopold (1949/1970), in *A Sand County Almanac*

> "Communities should always be imagined as in relationship to others, particularly downstream communities, rather than as singular and self-sufficient. An ecological re-conception of dwelling has to include a justice perspective and be able to recognise the shadow places, not just the ones we love, admire or find nice to look at."
> – Val Plumwood (2010), in *Shadow Places and the Politics of Dwelling*

In his book, *Ecolinguistics: Language, Ecology and the Stories We Live By*, ecolinguist Arran Stibbe (2015) points to these two quotes – one from the work of conservation ecologist Aldo Leopold and the other from environmental philosopher Val Plumwood – as examples of writers deploying the ecolinguistic tool of 'reminding.'

Stibbe defines reminding as "explicitly calling attention to the erasure of an important area of life in a particular text or discourse and demanding that it be brought back into consideration."

For example, a sustainable and ethical relationship with the natural world can only blossom when that world becomes available to our human senses: dirt we can touch, birds we can hear, trees we can smell. This suggests that loving a place, and therefore developing the desire to protect it from destruction, depends on our direct lived experience with that place. So, Leopold believed we can't protect something we don't love, and we can't love something we don't have direct experience with.

However, Plumwood's (2008) notion of 'shadow places' leads us into a paradox. As she explains, loving a particular place on the Earth can often mean loving that place to the exclusion of other places. This is because our love of place does not happen in a social and historical vacuum. Love of place always 'takes place' in a particular historical, cultural, and economic context. Today, within the dominant global economy we live under, our love for some places may unwittingly involve the sacrifice of many other 'shadow places' it depends on.

Shadow places, writes Plumwood (2008), are "The places that take our pollution and dangerous waste, exhaust their fertility or destroy their indigenous or nonhuman populations in producing our food ... the places ruined by and for fossil fuel production ...". In other words, shadow places are places designated as sacrifice zones to support the well-being of certain other places, and the people who inhabit those places.

There's a distinction we can make when writing about love of nature and place. On the one hand, there is a love of 'one's own place' that erases its connection to the vast web of other human and nonhuman places it depends on for its well-being. On the other hand, there is a love of place that doesn't erase those connections but actively works to *remind* us of them. Plumwood calls this latter approach 'the place principle of environmental justice':

> An important part of the environmental project can then be reformulated as a place principle of environmental justice, an injunction to cherish and care for your places, but without in the process destroying or degrading any other places, where 'other places' includes other human places, but also other species' places.

Tactics

Environmental Keywords

"[An environmental keyword is] a mode of worlding, a space of possibility for understanding and enacting worlds, with all of the opportunities and the limitations that this implies ... keywords become concrete foci around which complexity can be rendered in some way intelligible."
– Thom van Dooren (2019), in *Wake of Crows: Living and Dying in Shared Worlds*

Environmental keywords will be important tools in helping us tell new stories about environmental problems, impacts, and solutions: improving stories we already tell, and also helping us tell the stories that go untold. Human-environment relationships involve a complex tangle of social, cultural, ecological, political, economic, technological, and scientific stories.

To make sense of this complexity and tell new stories that can heal damaging human relationships with the natural environment, writers, scholars, and scientists are contributing to a new toolkit of environmental keywords for our time, to tell more powerful environmental stories that convince, transform and compel people to take action to improve human relationships with the more-than-human world.

The original inspiration for developing a toolkit of environmental keywords traces back to Raymond Williams' (1976/2015) *Keywords: A Vocabulary of Culture and Society*. At the end of his book, Williams left several blank pages to be filled in, instructing readers that his vocabulary "remains open:" "the author will welcome all amendments, corrections, and additions."

Now more than ever, we need to build a toolkit of environmental keywords to name and face our monumental environmental challenges. But what keywords might these include?

To give an example, consider the environmental keyword 'inheritance.' We live in a time of mass extinction: extinction of natures, cultures, and languages. We live "in a time of ongoing colonization, of diverse human and nonhuman lives," writes environmental philosopher Thom van Dooren (2019). In his research on conservation efforts striving to save the 'alalā, the endangered Hawaiian crow, van Dooren explains, "... taking care is always a historical and relational proposition: if we're doing it right, care always thrusts us into an encounter with ghosts, our own and others." These 'ghosts' might be evolutionary ghosts, like plants still waiting for seed dispersers now extinct to visit them. Or they might be historical ghosts, like Indigenous Hawaiians haunted by the illegal annexation of Hawai'i in 1893.

For 'alalā, these ghosts also include their language, a complex vocal repertoire of calls and songs lost over time as generation after generation await to leave their rehabilitation in captivity for a place in the wild. To understand all that has been lost, and what might still be saved, in this ongoing project of caring for 'alalā, van Dooren points to the work of Deborah Bird Rose (2004), who calls these uncertain projects of learning to inherit multispecies entanglements well, like the one to bring back 'alalā in Hawai'i, "the work of recuperation."

How might we do the work of ecocultural recuperation by inheriting the world well?

Hawaiian writer, Bryan Kamaoli Kuwada (2015), powerfully describes how Hawaiians navigate this question of inheritance in his essay, "We live in the future: Come join us." As he writes,

> Our genealogies are a backbone stretching to the very inception of these islands, and when we understand our genealogy, we know our origins, where we have been. We always have our ancestors at our back. That certainty gives us a wider possibility of movement, a more supple way to navigate through the world. (para. 8)

How might we better inherit the hauntings of extinction, colonial violence, and cultural loss that both 'violently elect us,' van Dooren asks, recalling the famous phrasing of Jacques Derrida, but that we also have the potential to transform?

Here's another question to consider: What kinds of histories have we inherited, whether chosen or forced on us – familial, emotional, cultural, social, political, economic, educational, colonial, ecological – and which inheritances might we recuperate to better navigate an uncertain future in more just and life-sustaining ways?

What other environmental keywords can help you tell your story?

Multiply the Perspectives

Environmental storytelling gets interesting when you multiply the perspectives or layer the different lenses that you bring to your story. No one perspective will give you the complete picture: only multiplying perspectives will give you the full-dimensional view of your topic.

So, when I first begin a writing project, one of the questions I always keep in mind is this: "What are the different perspectives – personal, cultural, ecological, historical, scientific – that I need to bring to bear on this subject?"

This tactic allows you to break your writing project down into separate sections or chapters, where each section delves into the topic you're exploring from a different perspective. For example, skilled environmental writers like Robin Wall Kimmerer (2013), Robert Macfarlane (2019), or Rachel Carson (1962/2002), all deploy this tactic in various ways by taking the reader on a journey across various perspectives. This allows them to layer depth and complexity by creating a narrative throughline from a particular vantage point. These different perspectives might include the view from neuroscience and psychology, recounting the phenomenology of one's lived experience – the memoirist part of ecowriting books – and of course, the important historical lens. As Pollan (2007) puts it: "History always illuminates things. How did we get here? Why did it take so long to get here? What did we learn along the way?". When it comes to writing about environmental issues, there are many ways to enlist the ecowriting tactic of multiplying perspectives. Some people zoom way out to look at international climate policy, while others zoom way in to bring nuance to technical discussions of carbon pricing or community debates around conservation programs to reintroduce endangered predators like wolves and bears.

As a discourse analyst of environmental issues, I think a discourse approach is relevant to all of these different ways of making sense of environmental issues. But my starting point is that the language we use matters in shaping how we discuss, understand, and therefore act on environmental problems.

When it comes to addressing the monumental environmental issues of our time – from climate change to species extinction – mapping out the dominant 'environmental discourses' circulating in the media and in our public conversations provides a shared system of meanings for us to define environmental problems, and therefore choose the best solutions to tackle them.

As the discourse analyst Norman Fairclough (2003) puts it:

> Discourses not only represent the world as it is (or rather is seen to be), they are also ... imaginaries, representing possible worlds that are different from the actual world, and tied in to projects to change the world in particular directions. (p. 24)

When asking about the ways we write about environmental issues, at the same time we should ask: what kinds of human relationships with the natural world do these ways of thinking, talking and acting help us imagine? In other words, what roadmaps for finding healthy pathways of living with other human beings and the more-than-human world do different environmental discourses, or perspectives, make possible?

Writing as an Ecosystem

When I encounter a roadblock in my writing, I find that a helpful writing prompt can work wonders in getting unstuck. Below, I outline one simple writing prompt that has worked wonders for helping me move forward when I struggle to string words and stories together.

I first learned about this tactic, "writing as an ecosystem," from an amazing course on nature writing offered by *Emergence Magazine* and taught by Chelsea Steinauer-Scudder.

Sometimes all that's needed to trigger your writing is a shift in perspective, to look at your subject from a different vantage point, as I mentioned above. This prompt helps you do just this by multiplying perspectives on a chosen topic. It allows you to layer memories, stories, and experiences on top of one another until a richer picture emerges of your subject, in all its many vivid dimensions.

The key approach at work here is the opening of yourself to a nonjudgmental process – serendipitous self-discovery. Fair warning: this method isn't a surefire tactic for writing success, as it has the potential to lead you into new territory as much as dead-ends:

> When you write, you lay out a line of words. The line of words is a miner's pick, a woodcarver's gouge, a surgeon's probe. You wield it, and it digs a path you follow. Soon you find yourself deep in new territory. Is it a dead end, or have you located the real subject? (Annie Dillard in *The Writing Life*, 1989 as cited by Chelsea Steinauer-Scudder in the *Emergence Nature Writing Course*).

Writing as an ecosystem:

1. Find an item of any kind: ideally, it should invoke an idea of nature or the environment, and is something small that you can hold in the palm of your hand. For example, it could be a weed, a leaf, a rock, a pinecone, a handful of sand, a shell, or perhaps a photograph of a plant, animal, or natural place.
2. Place the item next to a blank piece of paper.
3. You can either write the name of your natural entity in the center of your paper, or alternatively, you can simply place the item itself at the center.

4. Then, begin pondering this item: what memories, ideas, feelings, sensations, and associations come to you as you think about it? Mindmap these ideas on your paper. Important: keep your mental editor at bay, giving your mind the space to wander into new territories for 20 minutes or so. As your imaginative wanderings unfold, connect the flow of your ideas with lines or arrows that link them together. After 15–20 minutes, look at your mindmap. This is your writing ecosystem.
5. Next, select a location in your ecosystem that is away from the center, at least two steps removed from your item. Then, using that location as your starting place, write for 5 minutes. There are no wrong responses here, the aim is simply to generate writing in dialogue with the subject at that location. Now, choose a different location in your ecosystem, and begin the dialogue again from there. Repeat this as many times as feels right to you.
6. Finally, beginning with the center of your web, write a reflection about your natural item for 15 minutes, bouncing between your previous writings of different locations, as well as the whole ecosystem you created. You should let your imagination run wild, and see where it takes you, whether it be into new territory or dead-ends.

The magic of this simple little prompt is in the creative friction generated from moving between your previous writings and the whole thought ecosystem you created.

As you gather the narrative threads among your previous writings on memories, associations, and ideas related to your natural item, your imagination will take you in new narrative and descriptive directions.

As a writing-generating tactic, writing as an ecosystem encourages you to evade the constant surveillance of your top-down internal editor, allowing your ideas to messily emerge, branch out and fuse with other ideas.

You might think of your mindmap with its natural item at the center like a massive Wood Wide Web: an underground fungi network in a symbiotic relationship with your chosen item.

Here, the cobweb of fine filaments around your item – associations, memories, quotes, sensations, and emotions – act as a kind of prosthesis to support and nourish the emerging stories you will eventually link together, 'thickening' your story's filaments as you go.

Writing as an ecosystem is just this process: steadily extending a network of meaning that expands outwards into the soils of your experience, which then acquire nutrients, floating them back through the network to nourish your developing ideas.

For me, 'writing as ecosystem' appeals to my nonlinear, 'foraging-style' of writing, as Ursula K. Le Guin puts it, where stories never finish but are left, perhaps to be returned to and extended later on with the help of a fresh set of new connections. As the anthropologist Anna Tsing (2015) explains in her wonderful book, *The Mushroom at the End of the World*:

> Ursula K. Le Guin argues stories of hunting and killing have allowed readers to imagine that individual heroism is the point of a story. Instead, she proposes that storytelling might pick up diverse things of meaning and value and gather them together, like a forager rather than a hunter waiting for the big kill. In this kind of storytelling, stories should never end, but rather lead to further stories. In the intellectual woodlands I have been trying to encourage, adventures lead to more adventures, and treasures lead to further treasures.

I imagine ecowriting to be something like a rhizome of stories, where new connections can grow at any point along the narrative, leading to ever-new 'adventures and treasures.'

Conclusion

> "If we don't recognize writing as magic, we tend to fall under its spell ... I'm not saying writing is bad. I'm saying writing is a magic, and only when we recognize it as such can we use it responsibly."
> – David Abram (2020)

David Abram is an American environmental philosopher and 'cultural ecologist' best known for his landmark book, *The Spell of the Sensuous: Perception and Language in a More-Than-Human World*. His book bridges phenomenology – the philosophical study of experience – with the anthropological field of cultural ecology. It's a book that has stuck with me over the years ever since I first read it, where each paragraph shifts your perceptions of the world little by little until you've arrived at a dramatically different understanding of the world, and your place in it, by the end.

At the time I first read the book as a new college student, the book felt like a psychological windshield wiper, helping to unblur my interconnection with what Abram calls the "more-than-human matrix of sensations and sensibilities." In this way, Abram's work is not so much about helping us see the world in a new way. It leaves a more profound mark on its reader than that. More so, it embodies a writing style that induces the feeling that our anthropocentric thinking, sedimented over centuries through our immersion in the written word, has made us forget our fundamental interconnection with the natural world.

In *The Spell of the Sensuous*, Abram develops a way of writing grounded in the ecology of perception: a field of study that seeks to repair our broken sensory and perceptual connections with the more-than-human world through a new kind of writing, connections that have been broken, Abram suggests a little paradoxically, by writing itself.

When I first read Abram's book, I understood his main argument to be that ever since its invention, the technologies of writing and literacy have enraptured our human senses, with the effect of plunging us into an increasingly human-centered world. Writing has us under its anthropocentric spell, leading us to look deeper into our own human machines – from books to smartphones – instead of learning to read the language of the Earth as oral cultures have for millennia.

In other words, when I first read this book, I viewed it as an unconventional self-help guide to 'unlearning' the power of literacy to narrow my sensorial perceptions of the natural world. But it turns out I was mistaken. Like many who also read the book, I later learned, I misunderstood Abram's argument. Instead, the point Abram was trying to make is this: writing is not bad (he is a writer after all), but writing *is* a powerful form of magic.

In a fascinating recent interview, David Abram not only delves into his research on the ecology of perception but also seeks to debunk a pervasive myth about his argument that only literacy can lead to human disconnection with nature.

For Abram, it all begins with research on the ecology of perception: the study of how our senses are enmeshed in a 'more-than-human matrix of sensations and sensibilities.' It's worth quoting him at length on this point:

> I think as a cultural ecologist, what I'm primarily known for is research, investigations, into the ecology of perception or the ecology of sensory experience; that is, the way the activity of our eyes, of our ears, of our tongue, our nostrils, functions to bind our separate nervous systems into the encompassing ecosystem, as though our animal senses actually work almost like a kind of glue binding our individual neural system into the wider ecology, the wider ecosystem. (2020)

One of Abram's main arguments is that the written word often operates on our psyche like a potent form of magic that pulls us more strongly into its grip from childhood onwards. For much of human history, the cultural evolution of the human species involved a rich oral tradition, where human speech and nonhuman speech were seen as inherently interconnected. But the written alphabet disrupted these interconnections.

As writing spread like a viral technology around the world, the voices of nature, our once oral cultures that were so familiar, steadily receded into the background until one day we woke up in a silent world, where nature, and even the planet itself, had become merely an inert stage for human dramas to unfold. In other words, we have lost our ancestral knowledge and experience of Nature in the Active Voice. When I first read his book, I assumed Abram was telling me that writing has robbed us of our more primary perceptions of the natural world, and therefore, that writing has diminished our capacity to orient ourselves in the natural world. In sum, the magic of writing is its ability to put us under its powerful 'spell,' pulling us deeper into a human-made world.

Fortunately, this isn't at all the argument Abram actually made in his book, even though he regrets that many have interpreted it that way. As he said in an interview,

> One of the most common misreadings of my work and of my research has been to say, 'Oh, Abram is suggesting that writing is bad and that the alphabet is the cause of all our problems.' This is a terrible misreading, because I'm a writer and I love the written word and I love to read, and I'm deeply given to the exquisite power of the written word to open wonders. I'm not at all claiming – and this is quite important – I'm not at all suggesting that writing is bad, but, rather, that writing is magic, and that the alphabet is a very potent form of magic, a very concentrated form of animism.

As a form of magic, the potential for writing to both hinder and promote human and ecological well-being depends, Abram suggests, on how we use it. And, like any technology, it can be used in both responsible and irresponsible ways. But to say that writing is magic, "a very concentrated form of animism" as he puts it, creates an interesting spin on what it means to write, and to be an ecowriter.

In my view, Abram's exploration of the magic of writing reveals how, like any technology, writing channels us into certain ways of being, ways of knowing, and ways of relating to each other and the more-than-human world. Writing doesn't just represent the world. Writing creates worlds: it is a 'worlding' technology.

Like entering into the current of a fast-moving river, writing has the potential to carry us into new identities, new knowledges, new ecocultural relationships, and new worlds.

But what exactly does it mean to say that writing is 'magic?' As Abram explains, If we don't recognize writing as a very potent magic – that is, as something that has much more than rational effects upon our experience – if we don't recognize it as a magic, we tend to fall under its spell. The word "spell" has that double meaning, both to cast a magic within the world and also simply to arrange the letters. But those two meanings were once one and the same, because to learn to read with this new magic was to cast a kind of spell upon our own senses.

David Abram's idea that writing is magic reminds me of what an earlier perceptual ecologist, the Estonian biologist Jakob von Uexküll, called Umwelt. Umwelt is a German word that literally means "surround-world" but Uexküll used it to talk about the "experienced environment": the environment perceived by a creature through its unique sensory adaptations to the world.

Interestingly, Uexküll also used another phrase to describe the ecology of perception that different flora and fauna inhabit: their "magical environments." For Uexküll, animals' 'magical environments' are like the magical world children experience in play when they imagine characters from a fairy tale to appear in their room, or when dogs see, smell and react to something that's invisible to human senses. Uexküll believed the investigation of non-human beings' magical environments would only be accomplished through a combination of rigorous science coupled with a multi-sensorial form of storytelling that reattunes our human senses to the many different plant and animal Umwelts that represent the world too. Only by bursting our anthropocentric bubble – our all-too-human Umwelt – can we begin to know how radically entangled we are with other beings, and the multitude of magical environments that we share with them in earthly existence.

What takeaway might this have for ecowriting? Here's one suggestion: we might begin looking for ways to embrace the task, as David Abram memorably puts it, of "taking up the written word, with all of its potency, and patiently, carefully, writing language back into the land ... Planting words, like seeds, under rocks and fallen logs – letting language take root, once again, in the earthen silence of shadow and bone and leaf."

As ecowriters, if we take seriously Abram's call to harness the magic of writing, we'll need to find ways to transform our readers' ecology of perception by crafting 'more-than-human' stories: stories of multispecies entanglements where humans aren't the only heroes. I envision ecowriting as a collection of approaches to environmental storytelling that aim to push back against what Ursula K. Le Guin (1996) calls 'killer stories,' and instead, find ways to tell 'life stories':

> It sometimes seems that that story is approaching its end. Lest there be no more telling of stories at all, some of us out here in the wild oats, amid the alien corn, think we'd better start telling another kind, which maybe people can go on with when the old one's finished. Maybe. The trouble is, we've all let ourselves become part of the killer story, and so we may get finished along with it. Hence

it is with a certain feeling of urgency that I seek the nature, subject, words of the other story, the untold one, the life story. (p. 152)

Further Reading

Below are some of my favorite books that offer helpful examples of the different forms ecowriting can take. I hope they inspire your own ecowriting.

Abram, D. (1996). *The spell of the sensuous: Perception and language in a more-than-human world*. Knopf Doubleday Publishing Group.

Demuth, B. (2019). *Floating coast: An environmental history of the Bering Strait*. W. W. Norton & Company.

Ingold, T. (2000). *The perception of the environment: Essays on livelihood, dwelling and skill*. Psychology Press.

Kimmerer, R. W. (2013). *Braiding sweetgrass: Indigenous wisdom, scientific knowledge and the teachings of plants*. Milkweed Editions.

Lee, J. J. (2020). *Two trees make a forest: In search of my family's past among Taiwan's mountains and coasts*. Penguin.

Lopez, B. (2001). *Arctic dreams: Imagination and desire in a northern landscape*. Vintage Books.

Ogden, L. (2011). *Swamplife: People, gators, and mangroves entangled in the everglades*. University of Minnesota Press.

Pollan, M. (2007). *Second nature: A gardener's education*. Open Road + Grove/Atlantic.

Savoy, L. (2015). *Trace: Memory, history, race, and the American landscape*. Catapult.

Tsing, A. (2015). *The mushroom at the end of the world: On the possibility of life in capitalist ruins*. Princeton University Press.

van Dooren, T. (2014). *Flight ways: Life and loss at the edge of extinction*. Columbia University Press.

References

Abram, D. (2010). *Becoming animal: An earthly cosmology*. Vintage Books.

Abram, D. (2020, July 20). The ecology of perception [Interview]. *Emergence Magazine*. https://emergencemagazine.org/interview/the-ecology-of-perception/.

Berry, W. (2010). *What are people for?: Essays*. Catapult.

Carson, R. (1962/2002). *Silent spring*. Houghton Mifflin Harcourt.

Fairclough, N. (2003). *Analyzing discourse: Textual analysis for social research*. Psychology Press.

Fassler, J. (2013, April 23). The Wendell Berry sentence that inspired Michael Pollan's food obsession. *The Atlantic*. https://www.theatlantic.com/entertainment/archive/2013/04/the-wendell-berry-sentence-that-inspired-michael-pollans-food-obsession/275209/.

Ghosh, A. (2018). *The great derangement: Climate change and the unthinkable*. Penguin UK.
Haynes, D. (2003). *Art lessons: Meditations on the creative life*. Basic Books.
Johnson, A. E., & Wilkinson, K. K. (2020). *All we can save: Truth, courage, and solutions for the climate crisis*. Random House Publishing Group.
Kimmerer, R. W. (2013). *Braiding sweetgrass: Indigenous wisdom, scientific knowledge and the teachings of plants*. Milkweed Editions.
Kuwada, B. K. (2015). *We live in the future. Come join us*. Heheiale. https://heheiale.com/2015/04/03/we-live-in-the-future-come-join-us/.
otfelty & Harold Fromm (Eds.), In Ecocriticism reader: Landmarks in literary ecology (pp. 149–154). University of Georgia Press.
Le Guin, U. K. (1996). The carrier bag theory of fiction. In C. Glotfelty & Harold Fromm (Eds.), In *Ecocriticism reader: Landmarks in literary ecology* (pp. 149–154). University of Georgia Press.
Leopold, A. (1949/1970). *A sand county almanac*. Ballantine.
Macfarlane, R. (2019). *Underland: A deep time journey*. Penguin UK.
Plumwood, V. (2008). Shadow places and the politics of dwelling. *Australian Humanities Review, 44*(2008), 139–150.
Plumwood, V. (2010). Nature in the active voice. *Climate Change and Philosophy: Transformational Possibilities, 2010*, 32–47.
Rose, D. B. (2004). *Reports from a wild country: Ethics for decolonisation*. University of New South Wales Press.
Stibbe, A. (2015). *Ecolinguistics: Language, ecology and the stories we live by*. Routledge.
Tsing, A. L. (2015). *The mushroom at the end of the world: On the possibility of life in capitalist ruins*. Princeton University Press.
van Dooren, T. (2019). *The wake of crows: Living and dying in shared worlds*. Columbia University Press.
van Dooren, T., & Rose, D. B. (2016). Lively ethography: Storying animist worlds. *Environmental Humanities, 8*(1), 77–94.
Williams, R. (1976/2015). *Keywords: A vocabulary of culture and society*. Oxford University Press.

3. Let's Story about Storying[1]

DENISE CHAPMAN

My eyes still well up with tears when I think of the moment, when I, as a newly appointed teacher, thought I broke a child. I was a literacy intervention specialist working one-on-one with an 8-year-old boy named Cooper. We were in a classroom sitting under flickering fluorescent lighting. He and I were perched on dark blue plastic chairs with shiny metal legs and our elbows were rooted into the long wooden table with our fingers spread across our flustered cheeks. The room was smaller than an ordinary classroom: it used to be an AV closet, and the stale cold air smelled like a mix of copy machine ink and stacks of books that no one wanted. My student was also an object that no one wanted; he was the lone student who could not be taught to read. He was used to the routine of being pulled out of class for direct instruction on letter-sound relationships, but when I asked him to do as the program instructed of me, I could see the threads of our relationship tighten and thin. As he wrote out a string of letter A's, as the program directed, and mumbled the /a/ sound, my student's thick glasses began to quickly fog as tears dropped from the rims. *I broke him* – these were the words in my head, and it hurt. All that I learned at university dissipated, and I could only think of one thing to do – story.

I put one hand over the paper and let the other fall onto the top of his rounded back and began to story. I stated his full name and said, "I want you to ball up that paper and Michael Jordan it into the garbage can, 'cause Miss Denise was wrong. Let me tell you a story my grandpa told me and his grandpa told him." After that day, we began each of our lessons with storying. It became a way of forming common ground, an equalizing with

---

[1] This chapter has been developed and expanded from a blog post entitled *Stories can help us fly* (https://emergingwritersfestival.org.au/stories-can-help-us-fly-with-dr-denise-chapman/) from the Melbourne Emerging Writers Festival 2020.

wordplay that brought up a spiral of questions, a dialogic storying that ebbed and flowed like that call-and-response in a Black Southern church, or the comedic *doin' the dozens* creative joking in the back of the bus with the track team. The energy from our interlocking storying that also questioned what we were doing, the so-what of it all, shifted my student's disposition. He sat up like he was in charge and our sessions became so engaging that I once overheard his peers who were watching him being pulled out of class say, "Gosh, I wish I had a Miss Denise."

While I felt that our time storying was helpful, I did worry that perhaps this way of working with him, a way of working that harped back to how my grandparents would keep our minds moving, might not be truly supportive of my student, nor the other students that I began to teach. A few months went by and I was called into the boss's office. I was certain that my boss could hear my heart galloping in my chest. He said to me, "I know you're not doing what you're supposed to be doing with your students." I swallowed hard and my eyes shifted to shadows of a swaying tree just outside the office's bay window. He continued, "I know you're not doing what you're supposed to be doing because your students have never tested so well." Then he asked me what I was doing, and while it was not the teaching approach he subscribed to, we agreed that the positive outcomes we were seeing in the children's formal test scores, and the positive dispositions of both the students and their parents, were what we all wanted.

Storytelling. How might one explain what storytelling is? How can I make meaningful connections with you, the reader, about the act of storying? How can I begin a conversation with you about the mesmerizing moments of metaphor, about the dizzying heights of plot with purpose, or about the contagiousness of character? How can I do as my grandpa did – gather an audience 'round a narrative – but in this case, how do I story about story with my words fixed to paper? How can I imbue the feeling of freedom afforded to those who allow themselves to open their arms wide, let their feet rise with the lifting of voice, and let storytelling help both the teller and the told to fly? As a storyteller and poet, I must begin with a poem, with a poetic story that speaks to the emotive, spiritual power of story and what it can help us tap into as a catalytic tool for change.

> *I've been carrying work from oppression on both hips,*
> *on shoulders sloped,*
> *With fingers purple and palms cracked,*
> *My eyes carry bags packed with time lost,*
> *Ear lobes droop from the waterfalls of warrants stating "work past due".*

> *My legs move forward like thousand-year-old oaks walking the dimly lit parts of city streets at night,*
> *to find their breath,*
> *to shake off their soggy roots*
> *while caring for sleeping butterflies wrapped in their long leaves.*
> *The weight of the subordinated*
> *Interlocked with*
> *The weight of seeking purpose*
> *Like Newton's cradle*
> *My steps swing me forward*
> *Heel to toe*
> *Heel to toe*
> *Heel to toe then*
> *Toe to toe*
> *With weight that tilts my head*
> *Makes my chin rise*
> *Just enough so that my eyes can focus on the horizons,*
> *Past and present,*
> *Layers of color*
> *Each layer – a story*
> *Stories that my hungry mind can feed on,*
> *Stories that wrap my anxiety into a deep sleep,*
> *Stories that fill-in, color-in the iridescent outlines of perspectives silenced by those sitting on thrones oiled with the earth's essence*
> *and slanted right but not right,*
> *Stories that give purpose, tills seeds of being seen.*
> *Stories are our source of energy,*
> *A spiritual*
> *A liminal feeling*
> *A transformative touch*
> *An imaginative flow*
> *A place,*
> *A space of questioning*
> *what we know.*

I still feel connected to the transformative energy that came from Amanda Gorman's sharing her inaugural poem, her calling out, her televised counter-narrative to the United States. It caused me to leap to my feet and applaud, but why? There seemed to be something familiar and familial in her story. Much like a song from childhood I had heard before, I felt drawn to it. In her subsequent interviews with news media, she shared her mantra.

> I am the daughter of Black writers
> who are descended from freedom fighters,
> who broke their chains and changed the world.
> They call me. (Obama, 2021, para. 13)

Gorman's words, her poetry and mantra, shifted my spine – upward. I was reminded of a way of storying that I grew up experiencing. A specific way of storytelling that allows one to testify to their experiences so that their storytelling calls others to come together in critical communion, a disentanglement of challenges and a spiritual supportive (re)presenting of that which we know. When I heard Gorman speak, I felt this sense of wanting to talk-back, as if her public reflective telling was a spark for more reflective telling, wondering, questioning of meaning and knowing.

> *I can hear the storytelling,*
> *Like a cotton candy machine revving up*
> *Its sweet granules poured into the spinning drum*
> *The story, its sweetness heats up*
> *Drawing eyes to its telling,*
> *The spinning of tale*
> *Oozes out reflections*
> *Sweet, soft connecting threads,*
> *Enveloped flosses of resistance,*
> *Connecting us,*
> *Like cotton candy at a community carnival.*

The thing about cotton candy is the thread you pull out of the sugar, wound around your finger like a sticky spider's silk, and you know you could weave it back together with another thread if you weren't so intent on eating it. So, each story is a thread, and with the kids I taught, their threads began to form a tapestry of lessons.

The foundational threads of my tapestry come from one of the most intimate and warmly memorable childhood moments I can recall – experiencing stories shared by both my parents and grandparents. My chest flutters, my breathing deepens, and memories flood my mind like the first morning light on a hazy day. I can still feel my mother's soft arms wrapped around me as she told stories about being her brother's Harlem superhero and the occasional battle to protect his honor from bullies mocking his deafness. I can still feel my brother and me elbowing each other as we jostled for the center stage of Ezra Jack Keats' (1962) picture book "The Snowy Day," confidently held in my father's hands. It was the first time I had seen a protagonist with rich mocha skin who was going about his everyday routine socially unimpeded and with a contagious joy for discovering new things without fear. I can still feel the flood of calm that came from hearing my grandmother storying around how putting a single straw from a broom in your hair was a remedy for hiccups. It was the first time I realized that a Black woman's storying words of wisdom could be medicine for what ails ya. And, I can still feel the disappointment when my grandfather would say, "And now ya know, Dat's how it go," signalling the end of a set of trickster folktales told to him by his father's father. As a young Black American girl, this storytelling fuelled us forward and gave black folks the opportunity to see themselves as free.

At the time, I did not realize that what I was experiencing was both magical and subversive. My brother and I were the first in our family to experience childhood post-Jim Crow era (the practice of legalized racial segregation in the United States), and so the stories shared by my parents and grandparents were a collective gift, a resource of strength and hope. As Audre Lorde (2017) shares, "We were never meant to survive" (p. 201), and so my family's stories were a plea for our survival once the teaching baton was passed from home to the school system built on White supremacy. The stories my family shared (re)presented and (re)imagined black people and other marginalized peoples as protagonists in order to disrupt the dominant discourse we would encounter at school. These narratives validated our humanity, shifting us from an enslaved people to people with wings who had the power to fly, like in Virginia Hamilton's (1985) book "The People Could Fly."

However, once my brother and I entered the school grounds for the first time, the figurative armor gifted to us by our family seemed to rust, as we found ourselves drowning in stories that almost always centered our teachers' experiences as middle-class, non-disabled, White women from the U.S. South. I felt distant from the stories presented in school, and this made learning to read hard. The lives of "Dick and Jane" and their dog "Spot," made me feel alone and separated from my peers whose lives were like our teachers.' Even the opportunity to dream up fictional worlds can feel hampered when one is not provided spaces that are liberating (Thomas, 2019).

When I think of stories from my home, from my family, I am reminded of bell hooks' (2015) discussion around the construction of a homeplace, a space where one could feel safe, gain affirmation from others, engendering a collective healing from injustice. Storytelling can be a homeplace, an opportunity for growth and development, a source where one can nurture one's spirit, a space of community resistance, as well as a location for collective restoration and self-recovery. As bell hooks' (2015) shared, homeplace is a "site where one could freely confront the issue of humanization" (p. 42). Storytelling is that shelter where a person can feel free to reflect and connect and build community by countering narratives learned from the dominant society.

As I have aged, I have found that the most enduring storytelling moments I have experienced seemed to be tied to community, the space, the place, the land, the soil, the earth – to nature. This makes sense, as a story is not a good story without a setting. Storytelling experiences, interlocked with land, figuratively covered in soil, remain green in one's mind. The seeds from these storytelling moments are passed on and poignantly planted into one's mind as a child, flowering and then re-seeding the minds of our children when we tell stories tied to land and community. A narrative tied to nature is a narrative with a capital N. Narratives tied to nature are a nod of acknowledgment and a knowing that is threaded to what we see in our everyday, a swirling of place with a feeling of home. When storying about my experience as a cross-country runner, the soil and sun are embedded in my words. My cheekbones shift and glow, when I tell my children of my experiences sprinting through curtains of cool showers and curdling puddles of leachate on the piedmont leading me to the top of the capped former landfill, known as Mount Trashmore, one of the highest points in the flat landscape of Virginia Beach. When I see the heat shimmer in-between rows of corn, I am transported to the smells of summer. The smells are intertwined with memories of climbing crabgrass on the cul-de-sac curbs and the nearby fields of ripe strawberries. I see my friends blowing seeds off of dandelion flowers, the seeds kissing our shiny knees that are dotted with pink oil moisturizer used by the team of professional hair-braiders. Our mothers hover in the humidity as I sit in the middle of a line of girls shifting on the cement curb as we get our hair braided in cornrows outside, the hairdo representing summer freedom – freedom to roll down hills of grass, freedom to swim without worry, freedom to roll out of bed and straight out to play. I feel the freedom of our braids being finished with beads and tightly scrunched tinfoil on the ends. Looking into the hand mirror, I see dandelion seeds entwined in my braids and my mind twists and swirls with stories my father shared about how enslaved West African women took rice and other seeds from their homeland and braided them into their hair, which served as an act of autonomy and resistance, literally safeguarding seeds and hiding their knowledge of the earth within their crowns (Rose, 2020). "In Yoruba culture, hair is of such significance that the earth itself is sometimes personified as a woman having her hair combed with farming hoes. Because hair is associated with spiritual well-being" (Dabiri, 2019, p. 34). With these stories that interlink land and community, I am reminded that the "traditional Yoruba concepts of time were cyclical, and of the belief that the 'past' is not necessarily dispensed with but is in fact in dialogue with the future" (Dabiri, 2019, p. 2). Storytelling then becomes a way of dialoguing with the future, a way to seed a path for us to link narrative and nature.

Storytelling as a site for a safe coming together, a collective transformative talk, is something that has a long history in communities experiencing oppression. Within the African American community, the beginnings of this type of storytelling were in the form of song. Nikki Giovanni (2018) postulates that song, specifically a collective moan, was a medium of connection for the enslaved packed shoulder to shoulder in the bowels of slave ships. This moan, from deep in the soul, represented a wave of voice that transcended the many languages spoken and put forth common feeling. The moan was a call to the collective to join in a wave of voice, of song, of mutual storying that acknowledged their feeling of sorrow and, in the telling, solidified their being.

Looking back, I now know that my brother and I needed the experience of stories that represented what Rudine Sims Bishop (1990) describes as *Mirrors, Windows,* and *Sliding Glass Doors*:

> When children cannot find themselves reflected in the books they read, or when the images they see are distorted, negative, or laughable, they learn a powerful lesson about how they are devalued in the society of which they are a part. (p. xi)

Vivian Paley (1997) shared that children "are passionate seekers of hidden identities and quickly respond to those who keep unravelling the endless possibilities" (p. 4). I wonder what it would have looked like had my teachers considered storytelling like Paley (1990), who shared "storytelling is play put into narrative form" (*Storytellers and Story Players*, para. 11). Perhaps then it might be asked why some of us were pushed to the tight, edgy corners of that play.

If my parents and grandparents were trying to support my brother and me as children in these current times of Black Lives Matter, #MeToo, Fridays for Future, WeNeedDiverseBooks, Water Protectors, Extinction Rebellion, and Disability Rights Advocacy, I believe that they would be taking action, advocating for change, and putting forward movements like #OwnVoices, #DisruptTexts, #CripLit, #DVpit, #DearFlintKids, #WednesdaysForWater, #Ecoracism, #BlackJoy and Tolerance.org. I can imagine my family members insisting that this complex issue of seeing self in stories and connecting with nature is the beginning step in preparing and arming our young children with the tools for making our future a more equitable and sustainable one. Our children deserve to see true mirrors of our society, mirrors that are representative, inclusive, and equitable.

The stories in children's literature and other media can help children build bridges and engage with people they may not encounter in their everyday lives, and find connections to work together when they might not share common ground. Author Michele Norris (2020) notes, "Words are this connective tissue that allow us to listen and to find each other." Words also connect us to nature and community, and I believe that stories that are representative of our society fuel us all forward toward social belonging and the liberation we all need.

Nearly two decades have gone by since I first stumbled upon the powerful impact that storying can have in teaching. I still find myself telling the story about my experience with Cooper, as it represents more than the weaving of storytelling into my teaching practice. The story of my storying with Cooper is a telling of story with self, a self-reflection on me as a teacher and what is meaningful to me. Each time I repeat this story, I add a thread to my tapestry of teaching that honors my ancestors, reaffirms my humanity, and nurtures the value of ways of being and thinking that counter hegemonic discourse. Storytelling allows me to weave threads of the heart into what I do as a teacher. Nikki Giovanni (2018) shared, "We who do words are doing what we do. We are not trying to get folks who are frightened of us to be calm around us. We are reminding folks who love us that this is a good thing" (p. xvi). Storytelling is a good thing. Storytelling disrupts and sets us free.

> *Thought they done forgot 'bout us*
> *Cause ya know we're not like 'em*
> *We struggle and*
> *Dem masks don't quite fit*
> *And at intersections we choke*
> *But we stay together, learnin'*
> *Bout self and dat structure 'round us*
> *Together we strong*
> *We free*
> *Now we gotta clap*
> *Clap loud*
> *Stomp and sing like birds*
> *Ready to fly*
> *So dem folk with power can hear*
> *And we can all be free*

I wrote this poem in the persona of my maternal grandfather whose son was born severely deaf in Harlem, New York in the 1950s. Gaining equitable educational opportunities for a child who was poor, Black, and deaf during that time in history was nearly impossible, and instead, my mother's family had to collectively band together to give him the support that was afforded to them. While much has changed and been greatly improved since the 1950s, in terms of the educational experiences for children with disabilities, there are still tightly interwoven assumptions held by many in the education systems across both the United States and Australia, where I currently reside, that have led to great inequities likened to the time just prior to the civil rights movement of the 1960s. Our tapestry needs more threads.

References

Bishop, R. S. (1990). Mirrors, windows, and sliding glass doors. *Perspectives: Choosing and Using Books for the Classroom*, 6(3), ix–xi. https://scenicregional.org/wp-content/uploads/2017/08/Mirrors-Windows-and-Sliding-Glass-Doors.pdf

Dabiri, E. (2019). *Don't touch my hair* [Kindle edition]. Penguin UK. Retrieved from Amazon.com.au

Giovanni, N. (2018). Foreword: Our first stories. In S. S. Oliver (Ed.), *Black ink: Literary legends on the peril, power, and pleasure of reading and writing* [Kindle edition]. Simon & Schuster.

Hamilton, V. (1985). *The people could fly: American Black folktales* (Vol. 1). Knopf Books for Young Readers.

hooks, b. (2015). *Yearning: Race, gender, and cultural politics* [Kindle edition]. Routledge.

Keats, E. (1962). *The snowy day*. Penguin.

Lorde, A. (2017). *Your silence will not protect you*. Silver Press.

Norris, M. (2020). *Interview with Michele Norris about "The Race Card Project" at USD* [Video]. YouTube. https://youtu.be/wb-7cUgPVlw

Obama, M. (2021, February 4). The renaissance is Black: 'Unity with purpose.' Amanda Gorman and Michelle Obama discuss art, identity and optimism. *Time*. https://time.com/5933596/amanda-gorman-michelle-obama-interview/

Paley, V. G. (1990). *The boy who would be a helicopter: The uses of storytelling in the classroom* [Kindle edition]. Harvard University Press.

Paley, V. G. (1997). *The girl with the brown crayon* [Kindle edition]. Harvard University Press.

Rose, S. (2020, April 5). *How slaves braided rice seeds into their hair & changed the world*. Blurred by Lines. https://blurredbylines.com/blog/west-african-slaves-rice-hair-maroon-french-guiana-colonialism/

Thomas, E. E. (2019). *The dark fantastic: Race and the imagination from Harry Potter to the Hunger Games*. New York University Press.

4. Beyond Nature's Children: Examining the Environmental, Cultural, and Political Influences that Inform Indigenous Perspectives and Stories about Nature and the Environment

MELISSA GREENE-BLYE

"*Mis Misa* is the tiny, yet powerful spirit that lives within *Akoo-Yet* (popularly known as California's Mount Shasta) and balances the earth with the universe and the universe with the earth," according to Darryl Wilson, PhD, (Ajumawi and Atsugewi) author, storyteller, and elder. Wilson (2010) further explains:

> Its assigned duty makes Akoo-Yet the most necessary of all of the mountains upon earth, for Mis Misa keeps the earth the proper distance from the sun and keeps everything in its proper place when Wonder and Power stir the universe with a giant invisible *ja-pilo-o* (canoe paddle). Mis Misa keeps the earth from wandering away from the rest of the universe. It maintains the proper seasons and the proper atmosphere for life to flourish as earth changes seasons on its journey around the sun. (p. 56)

Similar notions of maintaining nature's balance can be found throughout traditional Indigenous spiritual beliefs with variations specific to particular nations' geographies and experiences, but with an emphasis on interconnection of humans as one part of a larger creation. This positioning is distinct from European-American spirituality, which tends to center or elevate humans as the pinnacle of God's creation. This distinction is crucial because it informs, not only the perception of nature's relationship to humanity, but also humanity's relationship and responsibility toward nature. This essay seeks to assist educators in their efforts to find authentic ways to tell stories about that relationship and responsibility from an Indigenous perspective. It is put forward in consultation with scholars and scientists whose lifework is dedicated to exploring and explaining Indigenous perspectives within their respective environmental or other scholarly expertise. It is also important to note that any work seeking to offer an Indigenous perspective must necessarily take into account the ways in which the United States' treatment of Indigenous peoples, past and present have made the issues of sovereignty and self-determination of paramount importance as they relate to not only environmental justice, but many other issues facing Indigenous peoples and nations today.

While the Ajumawi and Atsugewi identify Akoo-Yet as *wiumjoji elam-ji-se-la* [living spirit place], Wilson (2010) argues Euro-Americans view Mount Shasta as a natural resource, property of the United States, with animals and timber that must be controlled and harvested:

> Neither the individuals of the American government nor the individuals of the corporate state "see" the thousands of life forms that are a part of that forest. They do not "see" the bacteria necessary to grow the forest, they do not "see" the animals and birds that are displaced or destroyed as the mountains are shaved clean of forests. They do not "see" the insects and the butterflies of the forest as an element in balance with the universe. The mountain is talked about as part of land-use plans, not only for the resources it provides, but for its ability to draw tourists via newly developed ski resorts and as a draw for outdoor enthusiasts, as a means of economic development. (p. 57)

It is not just Indigenous sacred spaces that are compromised in the face of land economics; traditional economic systems are also often damaged as a result of large-scale resource extraction and energy projects on Indigenous lands. Economist, activist, and co-founder of Honor the Earth, Winona LaDuke (Ojibwe White Earth Reservation) (2010), argues that subsistence lifestyles are important sources of economic wealth and household necessities on reservations, but these Indigenous economic systems are not often recognized by non-Native industrial systems which results in negative outcomes for Indigenous populations:

> Resource extraction plans or energy mega-projects proposed for Indigenous lands do not consider the significance of these economic systems, nor their value for the future. A direct consequence is that environmentally destructive development programs ensue, many times foreclosing the opportunity to continue the lower scale, intergenerational economic practices which had been underway in the Native community. (p. 377)

LaDuke offers an eye-opening examination of the ways in which extraction and processing of resources such as uranium, coal, oil, and timber negatively impacted traditional reservation economies, with long-term consequences for the economic and physical health of reservation populations. LaDuke, alongside Indigenous environmental experts and scientists, holds that what is happening in their communities offers a micro-view of a much larger environmental crisis. This clash of economic and environmental perspectives was felt early-on by Alaska's Indigenous populations; as journalist Steve Talbot (2010) explains,

> Subsistence in the Indigenous world means more than the minimum necessity to support life. It means living close to nature, growing or gathering one's food, or hunting and fishing, with little reliance on a cash economy. Of course, a Native person must own or have legal access to the land and waters on which subsistence activities take place. (p. 389)

Talbot examines the way that the discovery of oil in 1968, and the subsequent passage of the 1971 Alaska Native Claims Settlement Act, diminished traditional land holdings and placed subsistence activities under state regulations without distinguishing between Native and non-Native interests, an important distinction according to Yupiaq scholar Oscar Kawagley (1995):

> Alaska Native peoples have traditionally tried to live in harmony with the world around them. This has required the construction of an intricate subsistence-based worldview, a complex way of life with specific cultural mandates regarding the ways in which the human being is to relate to other human relatives and the natural and spiritual worlds ... Native peoples developed many rituals and ceremonies with respect to motherhood and child rearing, care of animals, hunting and trapping practices, and related ceremonies for maintaining balance between the human, natural, and spiritual realms. (p. 8)

There would be ongoing tensions and negotiations in Alaska and across Indian Country over the course of the next several decades, carrying into the current moment. One of the most recent and memorable, due in part to the extensive coverage by non-Native media, involved the Dakota Access Pipeline project. Much of that coverage highlighted the oppositional elements instead of the proactive ones, and emphasized the objection to the pipeline rather than the ongoing struggle, led by Indigenous peoples, to protect their sovereignty and vital water resources. Historian Louis Warren (2004) maintains these tensions are not new and argues the necessity of examining Indigenous perspectives on nature by understanding the interdependent environmental, cultural, and political factors that shaped North American ecosystems in the nineteenth century and whose legacies carry into the current moment:

> In the twenty-first century, issues of land tenure and allotment, energy development and the "proper" use of tribal homelands and places, are only a few of the

> ways historians can see the continuing interaction between Indians and the natural world. Rather than showing us how far Indians have moved "away" from nature, such modern case studies remind us all the more of the many ways that Indian history and the history of contact with Europeans and Euro-Americans is inscribed on the land itself. (p. 303)

Warren asserts it is imperative that we move beyond popular culture portrayals and assumptions which rely on tropes that portray Indigenous peoples as "in harmony with nature" or as "the first environmentalists," making the case that this reinforces the stereotype of the "noble savage" and denies Indigenous peoples' agency in the myriad ways they proactively managed and cultivated natural resources for agricultural and community purposes long before Europeans arrived on the continent.

> By exploring how Indian people have perceived the earth, how they shaped and changed its natural systems, and how those changes in nature required changes in economy and culture, environmental history undermines the simpler myths of Indians as 'nature's children.' While demonstrating Indian power, innovation, and ingenuity, it also sheds new light on Indian conquest and its ambiguities. (Warren, 2004, p. 287)

It is important to acknowledge and respect the spiritual underpinnings that have long been a part of Indigenous storytelling traditions about nature, and also not let what we believe we know about that aspect of the subject based on a Western perspective lead us toward misperceptions and misinformation regarding use and conservation of natural resources (Brightman, 1993). Such an error also compounds the danger of leaving Indigenous peoples relegated to the past, failing to recognize their existence in the current moment, thereby negating them as stakeholders in current environmental issues.

Another important consideration in the ways we communicate and educate about Indigenous peoples and the environment involves the issue of naming. There is a lack of knowledge and consensus on authentic terminology for the landscapes and processes used by Indigenous populations for collecting, cultivating, hunting, and gathering. This begs the importance of finding stories and studies written by Native authors and scholars, with one note of caution, and that is the danger of assuming that all Indigenous peoples share a singular perspective on nature or their relationship with the environment, or that those perspectives have not been altered by younger generations (Wildcat, 2009). A sense of place, a connection to the land is an important part of continuity for Indigenous nations and an important underpinning of self-determination, so it is important to understand how government relocation and environmental management programs have factored into differences of opinion in many tribal communities, particularly since the creation of the Council of Energy Resource Tribes (CERT) in 1975. While CERT, a consortium representing multiple tribal nations, has increased control over energy resources on tribal lands, it has, at times, also been caught between differing opinions on the best way to utilize and/or protect those resources. Examining the connections between Indigenous peoples and nature is an ongoing and important area of scholarship, "In the twenty-first century, issues of land tenure and allotment, energy development and the 'proper' use of tribal homelands and places, are only a few of the ways [scholars] can see the continuing interaction between Indians and the natural world" (Warren, 2004, p. 303).

There are many stories to be told on both sides of those issues, including the story of how The Pawnee Seed Preservation Project (https://www.buffalosfire.com/video/pawnee-seed-preservation-project-seed-knowledge/) is working to revitalize knowledge and a seed bank so that Pawnee corn, once abundant in Pawnee traditional homelands (present-day Nebraska), can once again be grown by the community now located in present-day Oklahoma. And the story of Anna Lee Rain Yellowhammer (Hunkpapa, Standing Rock Sioux), a 13-year-old who started a petition to stop the Dakota Access Pipeline and participated in a relay to personally deliver the petition to leaders in Washington, DC (Charleyboy & Leatherdale, 2020). For younger readers, a plethora of stories can be found by searching the American Indians in Children's Literature (AICL) website (https://americanindiansinchildrensliterature.blogspot.com/). Some titles recommended by AICL Founder, Debbie Reese, PhD (Nambé Pueblo), include: *We are Water Protectors*, which was inspired by Indigenous-led movements and is also a call to action to protect water resources; *Nibi Emosaawdang (The Water Walker)* written in both Ojibwa and English; and *Nibi is Water (nibi aawon nbiish)*, a board book for toddlers written in English and Anishinaabemowih. For older readers, instructors might consider *Braiding Sweetgrass: Indigenous Wisdom, Scientific Knowledge and the Teachings of Plants* by Robin Wall Kimmerer, Ph.D. (Citizen Potawatomi Nation). Kimmerer, founder and director of the SUNY Center for Native Peoples and the Environment (CNPE), draws on her experience as an Indigenous scientist to explain that plants and animals are our oldest teachers and to examine how hearing their voices is key to engaging in our reciprocal relationship with the rest of the living world. The CNPE website (https://www.esf.edu/nativepeoples/) is a great starting point to explore work being done by and with Indigenous scientists and communities to offer research opportunities that seek to "evaluate educational strategies for bridging between western scientific and indigenous approaches in environmental disciplines." Another program worth exploring is the Haskell Environmental

Research Studies (HERS) program (http://www.hersinstitute.org/). Based at Haskell Indian Nations University, HERS is led by Indigenous scientists; this program seeks to "prepare tribal college students for graduate school and to help meet the challenges of climate and environmental change."

There is a great deal of work being done in the field, in the lab, and in the classroom to educate and encourage young people to engage with environmental issues and challenges. There is much to be gained when we can successfully combine current scientific knowledge with traditional, Indigenous perspectives on our relationship with the natural world. This is perhaps best encapsulated in the words of Jose Barreiro (2010), an Odawa elder, "The world is alive. Everything lives, including the stones and mountains. What makes us see this as one people, whether it is called 'Indian' or not, is that our elders understood about who the human being is in this world" (p. 478).

References

Barreiro, J. (2010). Visions in Geneva: The dream of the earth. In S. Lobo, T. L. Morris, & S. Talbot (Eds.), *Native American voices: A reader* (3rd ed., pp. 476–478). Prentice Hall.

Brightman, R. (1993). *Grateful prey: Rock Cree human-animal relationships.* University of California Press.

Charleyboy, L., & Leatherdale, M. B. (Eds.). (2020). *Not your princess: Voices of Native American women.* Annick Press.

Kawagley, A. O. (1995). *A Yupiaq worldview: A pathway to ecology and spirit.* Waveland Press.

LaDuke, W. (2010). Indigenous environmental perspectives: A North American primer. In S. Lobo, T. L. Morris, & S. Talbot (Eds.), *Native American voices: A reader* (3rd ed., pp. 376–388). Prentice Hall.

Talbot, S. (2010). Alaska natives struggle for subsistence rights. In S. Lobo, T. L. Morris, & S. Talbot (Eds.), *Native American voices: A reader* (3rd ed., pp. 389–395). Prentice Hall.

Warren, L. S. (2004). The nature of conquest: Indians, Americans, and environmental history. In P. J. Deloria & N. Salisbury (Eds.), *A companion to American Indian history* (pp. 287–306). Blackwell Publishing.

Wildcat, D. (2009). *Red alert!: Saving the planet with Indigenous knowledge.* Fulcrum Publishing.

Wilson, D. B. (2010). Mis Mis: The power within *Akoo-Yet* that protects the world. Originally printed in *News from Native California* (Spring, 1992). In S. Lobo, T. L. Morris, & S. Talbot (Eds.), *Native American voices: A reader* (3rd ed., pp. 56–70). Prentice Hall.

Websites

American Indians in Children's Literature: https://americanindiansinchildrensliterature.blogspot.com/
Center for Native Peoples and the Environment: https://www.esf.edu/nativepeoples/
Haskell Environmental Research Studies: https://www.hersinstitute.org/
Pawnee Seed Preservation Project: https://www.buffalosfire.com/video/pawnee-seed-preservation-project-seed-knowledge/

Part II: Teaching Ecowriting

5. Finding a Place in the World: Ecowriting with Elementary and Middle School Students

CINDY JENSON-ELLIOTT

Dry leaves crackle as I lead a gaggle of 15 seventh graders into a patch of riparian woodland. We're in a canyon a half-mile from school. We've gone over instructions; they know the drill: Find a spot 10 feet from any other student and sit down to write in silence. Their instructions are to use their senses to get to know this little bit of nearby nature and to write about it. If they want, they can write a haiku. Even though this is science class, we've read poems from Basho's (1966) *The Narrow Road to the Deep North*, and they know the structure and feeling of a haiku.

We've been down in the canyon before, looking for evidence of the ecological relationships we are learning about in science class. But today is different. Today we are here to get to know the canyon in a different way. As always, I am not sure how it will go. With seventh graders, you never know. Will they write? Will they get into the experience, or will they descend into seventh-grade silliness?

I find my own spot, set a timer and begin to write in my own notebook. It's hard for me to settle down. I'm on high alert for trouble. As a teacher, I'm always thinking of outcomes, of what-ifs, of the worst possible scenario. But now, in this singular moment in this specific place, some magic takes hold. Peace reigns. All around me, folded back on logs and stumps and piles of leaves, the students are silent, their pencils scrabbling across the page with a gentle scratch. For all the noise they are making, I could be in the canyon alone.

At the end of 15 minutes – an eternity for a middle schooler – we gather together and I invite students to share what they wrote. A few do, and we get to hear haiku that are both heartwarming and hilarious. For a moment, we are enveloped in the magic of place, the wonder of nature. Then, the roar of a distant car breaks the spell. We leap to our feet and walk back to campus, feeling connected to the canyon and to each other. We have just completed our first ecowriting challenge, and I can't wait to see what they write next.

Ecowriting with Elementary and Middle Schoolers

If we learned anything in our year of COVID-19 isolation, it's that kids need to feel connected with each other and with the planet in order to learn, and they need connection to lead emotionally healthy lives. Ecowriting is all about connection – connecting with the natural world and with other people. It is a way to connect Next Generation Science Standards (NGSS) concepts with real-life experiences. And it is a way to connect knowledge across the curriculum, from science, to social studies, to language arts and fine arts. Ecowriting opportunities are everywhere – inside the classroom, out in the schoolyard, in nearby nature, and even at home and abroad through remote learning. All it takes is a little idea, a notebook to write in, and a willingness to take some risks.

Keeping a Notebook

In whatever class I am teaching, whether it is in a science class, in a school garden, or in a writing lesson, we begin most days with low-stakes, ungraded writing in a notebook. While many students prefer digital notebooks, and I use these in the classroom as well, computers are hard to carry around outside, and real paper and pencil writing gives kids a way to connect ideas to reality through their hands. Writing daily – or close to it – is essential. Kids who write every day rarely fall into the false trap of perfectionism and writers' block. None of us write well every day, so whether your internal critic likes what you're writing doesn't matter. The point of writing is to get ideas down on paper or in a digital format, so that they can be shared. When they are writing in science, for example, they know I am only interested in their thoughts and ideas about the phenomenon we are witnessing. Notebooks are never graded on the quality of writing, but are examined as formative assessments to determine what students understand about an issue or idea. Writing in a physical notebook also gives kids a chance to draw, an effective way to develop ideas on the way to writing words.

Ecowriting in the Elementary School Garden and School Yard

Getting kids outside into the schoolyard or school garden with notebook in hand is the best way to begin a regular practice of ecowriting with elementary students. Even urban schools with very little outdoor space can find some way to connect with the outdoors. No plants? Chances are there is a weed popping up somewhere nearby. No garden? Somewhere on campus you'll find some dirt to dig into. No classroom animals? What animals are walking around in your schoolyard – ants? Pigeons? Pillbugs? Each living thing is an opportunity to connect students with nature. Each abiotic, or nonliving, part of the environment is a building block to helping kids see that the planet was here before us and will be here after we leave. There's security in knowing that.

Every part of the ecosystems of our schoolyards is a seed waiting to produce questions. How did that weed get there? Who is living in that hole in the ground? Where did that little pill bug come from? What are the ants doing? Why are those pigeons doing that weird dance? Ecowriting in the outdoors is a splash of water on a muggy day. It wakes us up – adults and kids – and gives a purpose to our work.

Weeds are phenomena of the first order, free plants just waiting to be used in a lesson on adaptations. In fact, I wrote my picture book *Weeds Find a Way* after discovering our school garden was full of weeds at the end of a spring break. We used them to learn how plants adapt to survive in a particular place by drawing them, learning about their parts, and seeing how roots, stems, leaves, flowers and seeds all work together to help the plant grow. These days I read the book to students before taking them outside to draw weeds and explore their adaptations, then bring students and weeds inside to create poetry about weeds.

Weeds Find a Way is far from the only picture book to help elementary students connect inside learning to outside ecowriting experiences. Every year more wonderful books hit the shelves that teachers can read inside before students begin exploring nature outside. Books such as *A Seed is Sleepy* (Aston, 2014), *Mama Built a Little Nest* (Ward, 2014), and *Be a Tree* (Gianferrari, 2020) help kids make connections between their own inside learning and the outside environment.

School gardens offer an opportunity for long-term ecowriting projects that blow apart boundaries of subjects and grade levels to involve reading, research, writing, and hands-on creation. As a long-time school garden teacher, I collaborated with teachers to create projects at all grade levels that incorporated planting in the garden with research, art and Making with a capital M – engineering bridges, creating informational signs and mosaic poetry pathways, even designing entire gardens. One-third grade project involved planting native medicinal plants, researching how our local tribes used the plants, and writing a guidebook to San Diego native plants. Two kindergarten fifth-grade buddy classes paired up to plant a butterfly garden, make butterfly stepping stones, and wrote a guidebook to the insects and spiders of our garden. Fifth-grade classes, learning about American history, planted dye plants and then dyed cloth. A multi-grade class of grades 1–3 learned about bridges, and designed and built a trellis strong enough to hold up a particularly heavy passionfruit vine before writing about bridges and what makes a structure strong. Each of these projects incorporated ecowriting as a way to bring inside literacy outside into the world in a meaningful way. For more ideas on incorporating ecowriting into the school garden at the elementary level, visit my blog from years of garden teaching: http://explorergarden.wordpress.com.

Ecowriting in Middle School: Writing to Save the World

Middle school students are notoriously disengaged from school. Developmentally, they are going through monumental changes in their lives, and social concerns often take precedent over academics. Ecowriting offers an opportunity to grab students' interest across the curriculum on public-facing projects that engage students by truly making a difference.

During the COVID-19 crisis, keeping kids engaged while learning at home was challenging, but it also presented unique opportunities to bring special guests to the classroom that many of us would never have considered otherwise. We participated in remote author visits, a tour of a weather plane with a National Oceanic and Atmospheric Administration (NOAA) hurricane hunter, and best of all, collaborative learning with classes in other parts of the country in a project I named the Climate Change Comic Project.

In the summer of 2020, I began planning a series of units for my sixth-grade students to study the atmosphere, weather, climate, and, at the end, climate change. To prepare myself, I attended many online climate change seminars for teachers around the country. At one, I found myself in a Zoom breakout room with Debbie, a teacher from Hawaii.

"Wouldn't it be cool," I said to her, "If we could collaborate on a project in which our classes could learn about climate change together?" I had an idea that had been floating around in my head for a while. Now that idea was coming together.

She was intrigued. "What do you have in mind?"

I explained my Climate Change Comic idea. First, two classes from different parts of the country and two different ecosystems would pair up and get to know each other. Each class would learn about the basic information of climate change on their own. Then each class would study how climate change is affecting their region by interviewing people whose work was affected by climate change, or whose work had an effect on climate change. They could learn comic drawing online from Alonso Nunez, a comic artist, teacher, and founder of Little Fish Comic Studio. Finally, each student would make a one-page comic about the person they interviewed, or about some aspect of climate change in their geographic region. Together, the classes would produce a comic book and sell it at an author book signing in which students could educate the public about what the issue looks like regionally.

Debbie was excited to join me on the project, but first I wanted to recruit teachers in higher latitudes. I began emailing a teacher in Nome, Alaska, trying to convince her to be a part of the project, and looking for another one in Maine. Ultimately, three classes across five time zones participated: a cohort of sixth and seventh graders at Deer Isle-Stonington Elementary in Deer Isle, Maine, a seventh-grade Sustainability class at 'Iolani School in Honolulu, Hawaii, and my school, Nativity Prep Academy, in San Diego. The students spanned not only time zones but also socio-economic boundaries, ranging from very low-income urban immigrant families in San Diego to very high-income urban families in Hawaii and to low- to moderate-income rural families in Maine. Each student's comic tackled a different issue or highlighted a different aspect of the problem. Students organized the comics by issue – ocean warming and rising, wildfire, wildlife, and the people working to make a change. The resulting comic book showed students how our common problems span the globe, how we have to work together to solve the issue. Best of all, the interviews with adults showed them how hard so many adults in such a variety of different professions are working to address and end climate change. While the project was about a difficult topic, it ended on an up note: that there is hope to be found in the efforts of many toward a common goal. For information and instructions on how to take part in the Climate Change Comic Project, visit my writing website, Words to Go, at https://www.wordstogosd.com/.

The Plastic Ocean Pollution Solution Project: Ecowriting and Public Speaking

While the Climate Change Comic Project focused on informational writing through comics, ecowriting can help teachers fulfill other aspects of the Common Core State Standards for reading and writing. Perhaps the most powerful use of ecowriting is persuasive writing with a real purpose: to convince powerful people in the community to make a needed change to environmental policy.

Twenty-two seventh-grade students in school uniforms sat in the front row of a public comments session of the San Diego City Council, waiting for their turn to present about plastic ocean pollution. Their presentation was the culmination of a months-long language arts project looking at the issue through reading, writing, hands-on projects, field trips, and public speaking. Now they waited quietly in their seats, listening while adult speakers took the floor to talk about issues that were important to them and their neighborhoods.

The students had spent months preparing for this moment. They had read about the harmful effects of plastic on the ocean, and learned what people and nations were doing to solve the problem. They designed and created a display board with information about the issue, focused on single-use plastics, and took it to the San Diego Maker Faire where they used their public speaking skills to teach the public how to sew their own utensil carriers using recycled upholstery sample fabric. They went on a field trip to the Scripps Institute of Oceanography, and then down to the tide pools to clean up plastics from the beach. They wrote persuasive essays about the issue and what they thought people should do about it, citing evidence from articles. And finally, they worked in groups to create short films about their own small, local solutions to plastic ocean pollution, and entered the films in a contest to apply to attend the Algalita Foundation's Plastic Ocean Pollution Solutions International Youth Summit in Dana Point, California. One team of three girls was invited to attend this pre-Covid event that gathered together students from the United States' mainland and Hawaii, Kenya, Tunisia, Indonesia, New Zealand, Mexico, Canada, the Bahamas, and a Pacific island nation. Their plan was to work on a project to persuade Mexican restaurants to cut their use of single-use plastics. However, once the conference was over, they moved on to pursue other interests.

Now, finally, it was the students' turn to present. One by one, they approached the podium, facing the full city council, while the Council Clerk shared their slides on a big screen. They each read a 20–30 second speech they had written for the occasion, each person explaining a different aspect of the problem of plastic ocean pollution, and in the end, asking the Council to ban the use of straws in our city. Afterward, some of the City Council members approached to thank them for the presentation and to pose for pictures.

Ecowriting as an Antidote to Despair

Ecowriting opportunities, whether they are as brief as a haiku or as long as a guidebook to native plants, offer elementary and middle school students an authentic way to build a relationship with the planet and with each other. They allow students a chance to make a real difference in a real, personal place. More than anything, they give students a voice in what goes on in their planetary home, a sense of efficacy and competence, an understanding that their words matter and can make the world a better place.

References

Aston, D. (2014). *A seed is sleepy* (S. Long, Illus.). Chronicle Books.
Basho, M. (1966). *The narrow road to the deep north and other travel sketches.* Penguin Books.
Gianferrari, M. (2020). *Be a tree* (F. Sala, Illus). Abrams Books for Young Readers.
Jenson-Elliott, C. (2014). *Weeds find a way* (C. Fisher, Illus.). Beach Lane Books.
Ward, J. (2014). *Mama built a little nest* (S. Jenkins, Illus.). Beach Lane Books.

6. Ecomedia Video Essays

Antonio López

How does *The Revenant* use cinematography to bring audiences closer to nature? How do films like *Into the Wild* construct the concept of wilderness? How do wildlife films warp time? Why do TV comedies get climate change so wrong? These are all examples of video essays – also called audiovisual essays – about media and the environment that are currently found on YouTube and Vimeo. A video essay is a concise video that visually explains a subject, presents a viewpoint, and builds a central argument by skillfully editing together video clips, audio elements, and images. Creating video essays helps students enhance their technical abilities while thoughtfully examining media content, constructing narratives, and using media-specific language. As they become adept in the language of visual media, the process of crafting video essays helps students understand and work with diverse modes of communication. Video essays also serve as a legitimate academic endeavor, allowing for a multifaceted evaluation of research, analytical thinking, and digital media proficiencies. It can serve as an accessible assignment for students lacking prior experience in media production, and educators without advanced technical know-how or extensive resources can incorporate them into their curriculum.

About six years ago, I began incorporating video essays as a core assignment in the senior capstone course I teach at an undergraduate university. The rationale at the time was that senior communications students should demonstrate their analytical and argumentative skills alongside an ability to utilize the aesthetic language of visual media. The assignment has been so successful that we have since adopted it across our program, offering it as an alternative or supplement to – but not in place of – traditional research papers. An unintended consequence is that it accommodates cognitive diversity. We have talented and intelligent students who struggle with traditional writing, but excel when allowed to express themselves through this form. Video essays strongly feature the student's voice, exemplifying how media analysis can be highly engaging and creative.

As the cliche goes, an image is worth a thousand words. Consider how advertising and social marketing deploy visual language to simplify complexity. Since environmental issues require a holistic, systemic approach, video essays can be useful for clear storytelling. Video essay projects can also incorporate a variety of learning outcomes and research goals, such as integrating information literacy assignments into the curriculum. This empowers students to enhance their research capabilities, critical analysis skills, and align with established institutional learning goals. Some examples include:

- Scrutinizing news coverage related to global heating and environmental justice movements.
- Investigating disinformation and fake news concerning the climate crisis.
- Applying critical thinking and deconstruction techniques to decipher environmental ideologies and rhetoric in advertising.
- Employing critical information literacy to evaluate the credibility of environmental claims presented in news media.
- Identifying instances of deceptive environmental assertions (greenwashing) in packaging, advertising, or news content.
- Examining the role of social media in either promoting or clouding discussions about the climate crisis.
- Addressing real-world environmental challenges and their corresponding solutions.
- Extending discussions about eco-ethics, rights, and responsibilities to encompass the living planet and workers.
- Establishing connections between the notion of the digital commons and environmental commons, such as air and water.
- Implementing alternative media practices to explore the dynamics of environmental change.
- Analyzing media organizations and their commitment to sustainability policies.

By leveraging video essays, educators can engage students in a comprehensive exploration of environmental issues while developing critical media literacy and research competencies.

This chapter's discussion of video essays draws on two perspectives. First, media literacy encompasses more than just applying critical thinking and analytical skills; it involves learning to effectively express and convey ideas through various media forms. Second, my specific focus lies in ecomedia literacy, which integrates environmental themes and concepts to promote cultural behaviors and attitudes aligned with eco-ethics. Within this context, students create ecomedia video essays that serve as a form of eco-citizenship, aiming to educate the public. To illustrate this, examples of student projects from my classes encompass a wide range of topics: An exploration of environmental themes within the *Lord of the Rings* film trilogy; an ecocritical analysis of Patagonia's surfer-oriented marketing strategies; examination of the *New York Times* coverage of the 2019 Amazon forest fires; analysis of the role of music in videos depicting the industrial slaughterhouse industry; establishment of principles for effectively communicating climate change visually; study of the visual language employed by vegan influencers on Instagram; critical assessment of the anti-SeaWorld documentary, *Blackfish*; delving into the representation of environmental ideology in *Fantastic Mr. Fox*; and an in-depth analysis of greenwashing strategies employed for iPhones. These student-created ecomedia video essays contribute to the wider goal of fostering eco-awareness and education, highlighting the potential for media literacy to inspire meaningful societal change.

In my role as an educator specializing in media studies and writing, I consistently emphasize that the foundation of all media creation rests upon the written word. Notably, the process of crafting video essays entails a substantial amount of writing prior to entering the production phase. Writing responsibilities span a spectrum, encompassing tasks such as drafting proposals, composing query letters, scripting narratives, and outlining treatments. In this chapter, I delve into the initial stages of video essay production, underscoring the significance of writing. This approach empowers students to explore in depth various environmental discourses, dissect rhetoric, and navigate the language intrinsic to audiovisual media. Ultimately, this activity enables students to effectively communicate insights related to the environment while also learning the art of proposal writing and becoming versed in the craft of audiovisual writing.

Video Essays: A Primer

A video essay is much like a written essay, but instead it is writing with video.[1] It presents a viewpoint supported by empirical evidence and theoretical

---

[1] Examples of video essays and links for resources can be found at: https://johncabot.libguides.com/communications/videoessay.

context, while leveraging the format and language of audiovisual media to articulate its arguments (Morrissey, 2015). Initially rooted in academic film studies, the popularity of video essays has surged notably due to platforms like YouTube and Vimeo (Bernstein, 2016).

Video essays encompass three core components: imagery (captured and found footage), sound (music and audio elements), and text (spoken and written content). This form is broadly categorized into four primary genres: supercuts, voiceover-based essays, text/image/sound-based essays, and desktop films. Supercuts amalgamate brief clips based on a specific theme, such as compiling scenes featuring trees from a particular movie or assembling instances of homophobic dialogue in a TV series like *Friends*. According to Filmscaple.com (n.d.), "Supercuts serve as both a fandom practice and a mode of audiovisual critique, revealing cinematic patterns, recurring themes in a filmmaker's oeuvre, and more." Voiceover-based video essays employ narration alongside audiovisual clips to elucidate and advance their analyses. Text/image/sound-based video essays forsake narration but integrate edited sequences of clips with textual elements to construct arguments. Desktop films provide a screen-centered experience, utilizing the computer's desktop as the primary stage. The screen becomes both the camera and the canvas for all content, including videos, applications, and programs typically viewed on a computer.

The production process for video essays unfolds through several stages: planning, gathering resources, editing, reviewing, and finally, presenting the finished work. In the following section, I focus primarily on the writing process for the planning stage, but I do briefly address some technical aspects of video essay production, which will make a vast difference in the final output. For starters, it is not necessary to have professional equipment or software, and there are free tools online to perform all the necessary tasks. You do not need to be a filmmaker or have a background in teaching video production to assign them, but be aware that common technical problems include poor audio (recording and mixing), low quality of source videos, over-reliance on text (also too much kinetic text or too many different fonts), and poor transitions.[2]

---

[2] In my experience, the most important but neglected element is sound. When recording narration (either using a camera, phone, or audio recording device), make sure it's done in an enclosed space without too much echo or hard surfaces that reflect sound. Use headphones to monitor audio while recording and don't touch the recording device with hands or mouth while recording. When recording audio, if a mistake is made, keep going— it's better not to start and stop the recording because quality will vary between recording sessions, which will be noticeable in the final mix. Mistakes can be edited out later. In terms of sound mixing, students tend to

The most common question people have is about copyright. Under fair use, we are allowed to use copyrighted media for the purpose of critique and commentary. I tell students to think of it the same way they would if they were writing papers. We are allowed to quote limited amounts of text in order to dialog with ideas or to perform analysis. Students must be careful to not appropriate the work of other video essayists. With so many video essays being made and accessible, it could be tempting to just copy someone else's original commentary, edits, and analysis. Extended sequences of someone else's original work should be credited on screen (if used at all). For example, the YouTube series, *Feminist Frequency*, has lengthy sequences of sexist representations in video games that were researched, assembled, and edited by the creator, Anita Sarkeesian. Incorporating those sequences would be bad practice because it's based on her original research and technical edits. Again, using the written essay analogy, if we inserted several pages of someone else's paper as our own without credit, it would be plagiarism and inappropriate. As a rule of thumb, the best practice is to create original sequences that represent the video essay's unique assertions and perspective. In terms of music, it's best to use royalty/copyright-free music (easily found on the web). The exception is if the music derives from the actual media being commented on (such as a film or TV show soundtrack).

To generate ideas about how to communicate about the environment with visual language, students can do the following warm-up activity.[3] The assignment is to create a 1-minute, single-shot mini-video documentary of something "natural." The video is shot on whatever device is available, such as a digital camera or phone. Students should scout place(s) to go and brainstorm what "nature"/"natural" means to them. The six rules are (1) no camera movement or in-camera action (pans, zooms, etc.); (2) no sound (mute mic or turn off for screening); (3) no editing; (4) no effects; (5) exactly 1-minute long; and (6) the topic is "nature" (anyway the student wants to interpret it). Screen the films in class and have a short discussion of how students interpret the intentions of the videomaker. This activity should help them prepare to think visually about their projects.

---

mix background music too high, making it difficult to hear narration. It's important to ensure there is enough separation between background audio and voice narration. In addition, try to avoid running a single soundtrack throughout the video without varying tempo or volume. Finally, don't mix audio with headphones, but use speakers (this is how most people will watch and hear it).

[3] This warm-up activity is adapted from Hadl (2016).

Writing the Ecomedia Video Essay

The initial planning phase of video essays encompasses the creation of a proposal, which I refer to as the "written video essay." This step involves students creating a written version of the project and introduces them to the valuable experience of crafting proposals, an underappreciated skill that few learn in their undergraduate studies. Importantly, students learn from proposal writing about being concise and avoiding verbosity. It's also essential for students to understand that robust writing is firmly rooted in research. Preceding the proposal stage, students should have already conducted preliminary research, typically in the form of a literature review or annotated bibliography. This research lays the foundation for generating ideas and formulating a thesis.

Writing for audiovisual media is a particular artform that has its own conventions. By following these guidelines, students can be introduced to the basic elements that make up this form of writing:

Title: The title serves as a preview of the video essay's content. It should encapsulate the core idea, striking a balance between being overly broad and excessively specific. Start by compiling keywords that encapsulate the central concept, then use these to craft the title. It should reflect a stance, i.e., "The Sound of Slaughter: The Music of YouTube Slaughterhouse Technology Videos."[4]

Logline: The logline condenses the primary concept into one to two sentences. It should utilize straightforward language, avoiding technical jargon. A robust logline employs a captivating "hook" to spark interest. It should encapsulate the primary takeaway and showcase an analytical viewpoint. For instance, "Patagonia promotes sustainable products and a more sustainable lifestyle to its clients. This video essay critically analyzes the brand's marketing from an environmental point-of-view." The logline could highlight any central conflict or offer a distinct insight. To practice, students can create loglines for existing films (ample examples are available online). Importantly, the logline is distinct from a subtitle.

Short synopsis: This concise paragraph, composed in the third-person present tense and active voice, functions as a snapshot of the video essay's content. Like a condensed version of the treatment (detailed later), it outlines what viewers will expect to see. It refrains from presenting arguments, justifications, or opinions beyond what audiences will view in the video essay, and doesn't delve into the thought process behind the making of it. To prompt the writing process, start with, "In this video we will see..." Another way to think about the synopsis is that it's like an abstract for a traditional research paper.

Motive: Direct students to be clear about why they are creating this by answering the three "whys": Why this topic? Why are you the one to do it? Why now? Also, have them account for additional aspects: the video's purpose

---

[4] Pro tip: It is advisable to re-write the title, logline, and synopsis after writing the treatment.

(inform, educate, motivate, persuade, entertain, enlighten, advocate, share, explain), the distinctive features of a video essay (in contrast to a written essay), and how it compares to existing video essays on the subject (if any).

Supporting research: Enumerate three to five sources (or as fitting for the assignment) from a previously prepared literature review or annotated bibliography. Accompany each with a one-sentence summary of the principal idea and why it applies to the current work.

Target audience and communication style: In a brief paragraph, summarize the intended audience and the suitable communication approach (serious, humorous, etc.).

Structure/Form/Style: Clarify whether the essay will adopt a voiceover, text/image/sound, supercut, or desktop film approach, and provide reasons for the choice.

Clips: Generally describe the kinds of footage, images, and audio intended for use, like "COVID-19 news clips from CNN," "Game of Thrones scenes exemplifying objectification of women," "screenshots of Trump's Twitter posts," etc. These form the data and evidence for the analysis, but do not require excessive details.

Technical issues: Students should identify potential obstacles they might encounter, such as obtaining materials or accessing equipment.

The next stage of writing involves producing a treatment and, optionally, a script:

Treatment: This written narrative details how the audience will experience the video essay by visualizing its content. Like the synopsis, the treatment is not a rationale for its creation but rather a description of what viewers will observe (show, don't tell). The reader should feel immersed in the project rather than reading a lecture on facts and intentions. Write with a present and active voice ("viewers witness" instead of "viewers will be witnessing") and abstain from exaggeration ("this exciting film explores," etc.). A treatment is not written like an essay, so it does not need to use conjunctive adverbs, such as "therefore," "moreover," "in addition," etc.

Script: Not all video essays entail narration, but if used, students should create a two-column documentary script with video on the left and audio on the right. The video column describes visual content, while the audio column contains narration text and any sounds viewers will hear. Templates are easily accessible online or you can craft a two-column table in your word processor.

Given that video essays predominantly rely on existing media as their source material, storyboards are generally unnecessary. They find greater relevance in original scripted projects, aiding in the visualization of action, camera angles, movement, and framing. An exception arises with desktop films, where the coordination of screen actions becomes pivotal and necessitates the use of a storyboard to choreograph sequences.

Assessment

My method of assessment is divided into two parts: formalistic elements and research. Obviously there will be a difference between the written and final versions of the video essay, but you can glean from the written version many formalistic elements. I assess: title (encapsulates main idea); take-home message (clear, concise); technique (edits, images, transitions); aesthetic choices (look, feel); creativity (pushes boundaries, unconventional); audio (clear, well enunciated, good mix); on-screen text (readable, grammar, spelling, typos, sentence structure); music (appropriate, copyright free); fair use; and length. For research, I assess: research design; theoretical framework; sources (diverse, academic); empirical evidence (demonstrates research); organization (good transitions, definitions, structured argument); clarity (thesis stated and supported with examples, evidence, background, context, take-home message); script (well-written); analysis (goes beyond description); understanding terms and grasp of concepts (definitions); originality of concept; style (flow, makes sense, creative, academic, command of English); and effort (passion and interest).

Conclusion

I consider video essays an important tool of ecomedia literacy. As I conceive it, ecomedia literacy promotes a paradigm based on systems thinking and an ecocentric belief in the interconnectedness of life, humans, technology, and economics grounded in eco-ethics (López, 2021). As a starting point, ecomedia literacy incorporates several learning objectives so that students learn to:

- Grasp the interconnected material relationship between media and living systems, examining their role in biodiversity loss, water and soil contamination, global heating, and worker health.
- Examine the interdependence of information communication technologies with the global economy and development models.
- Investigate how the current globalization model correlates with the history of colonialism, impacting living systems and ecojustice.
- Differentiate between anthropocentric and ecocentric discourses.
- Analyze how media establish symbolic connections and narratives that advocate environmental ideologies and ethics.
- Assess the phenomenological impact (affect) of media on perceptions of time, space, and place.
- Recognize and critically engage with the inherent epistemological bias of modernity.

- Foster an understanding of the ecomedia commons.
- Apply principles of eco-ethics and eco-citizenship to actively respond to the challenges posed by the climate crisis.

Combined with the video essay, there is a range of strategies for attaining these ecomedia literacy objectives. They include utilizing reverse curriculum design founded on problem-solving and solution-oriented results, constructing diverse future scenarios, challenging conventional human-nature divisions, shifting from theoretical knowledge to hands-on learning linked to nearby ecosystems, transitioning to a political ecology framework rooted in ecological economics, and reshaping ecological metaphors to prompt learners to recognize media as integral to the living planet. These approaches collectively strive to restore and advance ecocentric perspectives.

In my role as an educator encompassing media creation and analysis, I firmly believe in their inseparability: crafting and dissecting media should ideally form a continuous cycle. Gaining insights into rhetoric and the nuances of media language also empowers students to enhance their communication skills significantly. Media education should mirror a dialectical process, involving enjoyment and critical engagement. Much like how learning to play a musical instrument or delving into music theory enriches our understanding of music, it also equips us to critically assess music industries, standardized formulas, subpar musicianship, and artistic limitations. When integrated with ecomedia literacy, video essays present an optimal avenue to accomplish precisely this level of critical engagement and pleasure.

References

Bernstein, P. (2016, May 3). What is a video essay? Creators grapple with a definition. *Filmmaker Magazine*. https://filmmakermagazine.com/98248-what-is-a-video-essay-creators-grapple-with-a-definition/.

Hadl, G. (2016). *Kankyou media riterashii (EcoMedia Literacy)*. Kwansei Gakuin University Press.

López, A. (2021). *Ecomedia literacy: Integrating ecology into media education*. Routledge.

Morrissey, K. (2015, September 20). Stop teaching software, start teaching software literacy. *Flow*. https://www.flowjournal.org/2015/09/stop-teaching-software-start-teaching-software-literacy/.

Supercut. (n.d.). https://www.filmscalpel.com/portfolio_page/supercut/.

7. Visual Journaling as Ecowriting

PEACHES HASH AND THERESA REDMOND

Introduction

Over a decade ago, Louv (2008) wrote in *Last Child in the Woods: Saving our Children from Nature-Deficit Disorder* that children and adolescents are aware of global threats to the environment, but their physical contact and intimacy with nature was "fading" (p. 1). This fading has been resulting in a disconnection from the natural world, which Louv (2008) explains is the "human cost of alienation from nature," including "diminished use of the senses, attention difficulties, and higher rates of physical and emotional illness" (p. 34). To reduce this rupture, Louv (2008) urges readers to attempt to heal the "broken bond between our young and nature," not only for reasons related to aesthetics and justice but also because "our mental, physical, and spiritual health depends upon it" (p. 3). Within a year of the publication of Louv's (2008) book, Morrison (2009) wrote that "our culture is living under a collective illusion – the belief that this planet is composed of a collection of unrelated and independent objects, rather than interrelated and interdependent subjects that make up a fragile and miraculous web of life" (p. 104). In the same edited collection, Buzzell (2009) expressed that "[o]ur present lifestyles seem to be designed for the convenience and benefit of the industrial machine culture we inhabit, not for our ancient body-mind-soul" (p. 49). Louv (2008), Morrison (2009), and Buzzell (2009) have all been concerned with how humans perceive and interact with nature, causing them to write as a call to action for readers. These scholars urge readers to explore ways humans can restore their personal relationships with the natural world, especially when interacting with young people. When examining this text today, we as writers and educators should consider what, if anything, has improved in education to reduce this disconnection. Have young people's relationships with nature gotten better or worse with the rise of social media and access to digital technologies, especially after a year of isolation due to the global pandemic caused by COVID-19?

This essay is a synthesis of our shared knowledge and dreams for education. We believe that ecowriting is a pedagogical tool for challenging the growing detachment from nature. Furthermore, visual journaling involves multimodal compositions that offer benefits for students of all ages and ability levels. As a form of expressive arts that connects to ecotherapy and ecoeducation, visual journaling unites students with the natural world in creative, pleasurable ways. In these and more ways, visual journaling is a "radically democratizing creative practice" (McCaughey & Redmond, 2021). Visual journaling may be enacted across content areas and throughout a student's life to invite students' voices, facilitate metacognition (thinking about one's thinking), and build relationships to knowledge and the world. Through visual journaling as a practice of storying, of expressive arts, and of ecowriting, we may access and cultivate new pathways and possibilities in teaching and learning about climate change and other issues of environmental justice in ways that deepen students' perceptions, perspectives-taking, compassion, and advocacy. As England and her colleagues (2019) write, "Storying climate change activates empathy, agency, and collective action – skills necessary for responding well to climate change" (p. 21).

Through this essay, we hope to show how visual journaling, as an ecowriting process, may disrupt the culture of alphabetic text as a dominant mode for learning. This can open an expressive practice for students of myriad backgrounds and experiences to make sense of the complexities of the climate crisis, as well as embody and embolden their relationships to the natural world. As former Visual Arts and English teachers who have worked across K-12 settings, we have both come to visual journaling as a teaching and learning pedagogy that is expansive in cultivating students' opportunities to engage in problem-finding and problem-solving through reflective, recursive, and generative ways.

Expressive Arts, Ecotherapy, and Ecoeducation

Expressive arts are practices of art-making that are used in therapeutic and educational settings to promote introspection, construction of knowledge, development, and reflection. Unlike studio-based art courses, expressive arts focus on the process of art-making over skilled products of art (Knill & Knill, 2017) and define an "artist" as anyone who puts time and effort into an artistic product (Shore, 2009). Expressive arts include all forms of art-making for expression, including the visual arts, music, movement, and written text. In educational settings, expressive arts can foster personal connections with materials, engage students through active learning (Hash, 2020b), create opportunities for expression (Hash, 2021), and promote inclusivity by providing ways for students to use their preferred methods of learning to display knowledge (Hash, 2020a). Ecotherapy and ecoeducation overlap with expressive arts when these practices involve art-making as a tool. Buzzell and Chalquist (2009) explain that ecotherapy is an "umbrella term for nature-based methods of physical and psychological healing" that "acknowledges the vital role of nature and addresses the human-nature relationship" (p. 18). Although ecoeducation and ecotherapy occur in different settings and are used by different types of practitioners, they both are, as Clinebell (1996) explains, "complementary healing and growth processes" (p. 62) that "first seek to enable people to accept their continual dependency on nature as crucial for the well-being of themselves and the earth" (p. 33). This personal connection to the natural world is vital for students to begin to understand the interconnectedness between their existence and all living things on this earth (p. 14). Through ecotherapy and ecoeducation, ecobonding, or "claiming and enjoying one's nurturing, energizing, life-enhancing connectedness with nature," and ecophilia, "the love of life associated with this bonding with the earth" (Clinebell, 1996, p. 26) are more likely to occur, thus planting foundations for ecoliteracy that could lead to ecojustice.

Ecoeducation and ecotherapy include nature-reconnection practices that may involve art-making (Buzzell & Chalquist, 2009). The goals of these practices are to:

- facilitate healing/alienation from the natural world and an openness to the ways nature affects the body, mind, and soul;
- help people become more aware of the lived experiences and learning that occur in nature; and
- motivate people to adopt "more earth-caring lifestyles and behaviors that will help save the biosphere" (Clinebell, 1996, p. 62).

Ecowriting meets these goals by creating opportunities for bonding creatively with nature. Writing is an artistic form of expression that can help students make personal connections to material and express their emotions. Moreover, multimodal forms of writing such as visual journaling can serve as a form of ecowriting that is more enjoyable, open, and expressive for students.

Visual Journaling

Visual journaling is the process of documenting, representing, finding, or (re)working thoughts, ideas, information, issues, questions, and other experiences graphically. A visual journal is like a sketchbook, diary, or art journal and generally incorporates images, words, and other fodder – such as newspaper clippings, magazine elements, art supplies, found materials, and natural matter – brought together in a physical notebook or journal. Engaging in visual journaling can be done with a range of traditional art supplies, such as acrylic paints, watercolors, washi tape, oil pastels, color pencils, and more. However, as an emancipatory practice, expensive supplies are not required and may even confine expressive potential. As Hutchinson (2018) explains, visual journaling "focuses our intention on our thoughts, feelings, and actions" (p. 2). During the shelter-in-place mandates of the 2020 COVID-19 pandemic, many art teachers and classroom teachers alike were prompted to think creatively about materials to which all students would have access. Woodard (2020) shares her uses of spices for "pantry paintings" to facilitate equity in expression for her students. Likewise, objects from the natural world – such as leaves, rocks, sticks, and feathers – make inviting instruments for mark-making or printing in one's visual journal.

In the classroom context, visual journals can be used with open-ended prompts, focused on particular themes or subject matter, or enacted via free journaling. Similarly, visual journals can be individual or shared, with a continuum of possibilities in between. For instance, we have included visual journaling where students begin working (or playing) in their own journal and then pass or exchange their journals for an extended activity or prompt. Visual journaling offers pedagogical benefits for students in K-12 classrooms, including taking students off "auto pilot" as they make connections between their experiences and feelings and academic content (Hutchinson, 2018, p. 1). Because images are more inclusive for comprehension and less restrictive than alphabetic text alone, visual journaling provides a way for students to express, reflect, and communicate with nuances and depth that words alone are less likely to achieve (Ganim & Fox, 1999). Additionally, visual journaling helps improve clarity, mindfulness, self-awareness, processing speed, and creativity (Hutchinson, 2018).

Ecowriting Through Visual Journaling

Through its (re)combination of images, words, and other materials, visual journaling may be conceived of as a multimedia form. Unlike a sketchbook or art journal, however, visual journaling is an organic and intuitive process that typically unfolds without preconceived design, heavy planning, or artistic training. In essence, "visual journaling is an expressive process that employs both images and words in order to inquire into and reflect on phenomena aesthetically" (Redmond et al., 2021, p. 125). Visual journaling and ecowriting are a natural pair, particularly as both invite the affective or emotive arenas of our being as vital in the creation process, even more so than artistic skill. Levine (2017) explains that engaging in aesthetic experiences in learning is fundamental to human experience and a "practice of the imagination that can lead to the discovery of the truth of the situation in which we find ourselves" (p. 179). In the context of the climate crisis and as a form of ecowriting, visual journaling may facilitate a "movement towards integration and wholeness" (p. 179). Similarly, it is a practice that benefits not only students, but also teachers and can be incorporated into professional development as well as curriculum design (Hash et al., 2022). Figure 7.1 features an example of a visual journaling spread created by the second author, Theresa, as she worked out how to incorporate a new ecomedia literacy unit into her undergraduate media literacy class. Specifically, she sought to develop students' awareness of mediated space as a construct separate from experiences in the natural world and to identify the physical impacts of being "plugged in" to anthropocentric framing. Notice the incorporation of both images and words alongside hand-drawn lines and found images, including houses and cards from repurposed wrapping paper.

Figure 7.1: Example of visual journaling as ecowriting for curriculum planning purposes

A great deal of writing in K-12 classrooms involves the use of screens. Creating writing assignments that involve nature are beneficial to students, but teachers should also consider *how* students are composing their information. In *Educating for Eco-Justice and Community*, Bowers (2001) asks readers to question if technology allows "for further development of the skill and insight of the user" (p. 169). While screens offer students ways to view natural elements that they cannot physically travel to, they also serve as barriers to embodied learning. Although multimodal forms of writing are often synonymous with students composing with screens, Shipka (2011) explains that when educators attempt to liberate students from the boundaries of the page with the use of screens, they institute another limitation to how texts can be "composed, received, and reviewed" (p. 11). The use of technology, then, results in "missing or undervaluing the meaning-making and learning potentials associated with the uptake and transformation of still other representational systems and technologies" (p. 11). Bowers (1995) believes that to educate for an ecologically sustainable culture, teachers must resist the dominant view of technology as always the most effective tool for learning, and instead create space for embodied intelligence where learning promotes relationships and creative problem-solving (p. 125). Visual journaling promotes embodied learning because it calls for active problem-solving and reflection that can be experimental, nonlinear, and reflective. It offers ways for students to play with materials, exploring nuances of their emotions and their experiences. Additionally, visual journaling can spark changes in behavior. The process of visual journaling helps people reflect on their experiences, the information they consume, and the tasks they perform (Hutchinson, 2018), providing a macro view of experiences that gives the opportunity to course-correct the trajectory of one's decisions.

The spirit of visual journaling is well-aligned with the purposes of ecowriting and, in many ways, visual journaling may be conceptualized as a form of ecowriting. Visual journaling invites the use of analog materials – including physical art supplies, found items, and natural materials – and encourages slow looking. Through visual journaling as a form of slow looking (Tishman, 2017, as cited in López, 2020), students are invited into a reflective and generative experience that is personally and relationally immersive, allowing one to process complex ideas and issues through intuitive or intentional expression. López (2020) explains "... introducing slow media to students is an important aspect of ecomedia literacy pedagogy" (p. 241). With the rapid proliferation of digital media during the COVID-19 pandemic in 2020 and the continued push to use digital technologies in schools that has followed, there is less space in the curriculum for slow looking and, thus, for the kind of deep reflection and learning that is vital for ecowriting. Visual journaling, as an ecowriting approach, not only facilitates slow looking, but also reduces reliance on the use of fast media through its invitation of simple art supplies and, even, found materials.

Applications for K-12 Classrooms

When outlining his curricular model for an ecologically sustainable culture, Bowers (1995) envisioned students engaging in activities that reveal how all things are connected: humans, the industrial environment, and the natural environment. Ecowriting can facilitate this connectedness, beginning with educators providing students ways to personally bond with nature. Students may experience alienation, fear, and disconnection from the natural world, and a method to ameliorate this in ecoeducation is to provide students opportunities to share their ecological story (Clinebell, 1996). Visual journaling as a form of ecowriting provides nonlinear, creative, and enjoyable ways for students to tell their ecological stories. When visual journaling, students construct knowledge, experiment with materials, and fuse aspects of their identities with course concepts. Visual journaling can be adapted for different content areas and ability levels, including:

Natural Sciences. In his book *Experience and Nature*, educational philosopher John Dewey (1925) identified a divide between personal experience and scientific inquiry. He envisioned a way for the two elements of learning to "get on harmoniously together," where experience serves as a method for learning about nature, and, in turn, this learning serves to deepen, enrich, and further an experience (p. 2). Dewey (1925) saw language as a way to facilitate this process because of its expressive quality, but language does not have to be restricted to alphabetic text. In fact, Dewey identified art as an effective practice for displaying knowledge of the natural world because it provides depth and pleasure. Visual journaling can be used for multimodal scientific notetaking, reflections on course concepts, and expressions of lived experiences.

English Language Arts. Louv (2008) viewed nature as an immersive experience for students that could awaken the senses and inspire creativity. In *Last Child in the Woods: Saving Our Children From Nature-Deficit Disorder*, he explains how being in nature heightens the senses, which brings increased awareness of one's environment. Visual journaling as ecowriting can serve as a place for students to record their sensory experiences in nature, reflect on aspects of their experiences, and compose their connections with the natural world.

Social Studies. Writing for the National Council for the Social Studies (NCSS), Sperry and Baker (2016) urge educators to include new media forms in their curriculum as a means to address today's digital age learners. They explain that by including "analysis and production into the social studies we expand our classrooms to include the modes of communication that dominate the lives of our students. This is particularly important for non-print oriented students who are sometimes alienated from their academic experience" (p. 185). Among their suggestions are ideas to include the climate crisis in social studies education. Through visual journaling, students can incorporate non-print based representations to process their social studies learning, including how issues related to the climate crisis are interconnected with social and economic structures and concerns.

Two Examples of Ecowriting Through Visual Journaling

López (2020) states, "an ecomedia literacy curriculum needs to be structured in such a way that sparks the curiosity of students to be engaged and interested in the particular relationship between media and ecology" (p. 239). A powerful pedagogical benefit of the arts is that they spark curiosity by disrupting typical classroom procedures and ways of knowing. Visual journaling is aptly incorporated into critical media literacy practices, including ecomedia literacy curricula, in myriad ways to ignite engagement and stimulate critical thinking about our perceptions of the climate crisis and our relationships to the planet. Along with the broader applications suggested above, we provide two specific examples for how to incorporate ecowriting through visual journaling to augment critical media literacy practices in learning. While these examples were each conducted in an undergraduate media literacy class, both may be adapted for K-12 contexts and aligned with specific learning goals. We offer ideas for how to adjust the activities following each summary.

Ecoliteracy Hands. In this ecowriting activity, students engaged in guided visual journaling following an introductory lecture that exposed them to the constructs of anthropocentrism and ecocentrism using the work of Antonio López (2014, 2020). Before the lecture, students were invited to trace their hands and paint either the space around or inside of the form using a palette of earth tones in green, blue, and brown. The hand functioned as a frame to represent the anthropocentric construction of nature through media and technology. Following the lecture, students were invited to revisit their painted pages and respond with words. The prompt provided was to simply share words, phrases, ideas, or prose that related to how they were processing the concepts covered in the lecture. Key concepts included: root metaphors, anthropocentric discourse, ecocentric discourse, planned obsolescence, perceived obsolescence, and ecomedia literacy. Figures 7.2, 7.3, and 7.4 provide

Visual Journaling as Ecowriting

Figure 7.2: Example 1 of students' ecoliteracy hands: "And how often do I forget my place, How often do I realize the extremes that my actions enact. No one gave me the right to be destructive. But I accept their invitation anyway."

three examples of students' visual journaling pages from this activity. Their reflections demonstrate what Clinebell (1996) explains are the ameliorating effects of ecoeducation, whereby students are able to process their alienation, fear, and disconnection in order to tell their own ecological story.

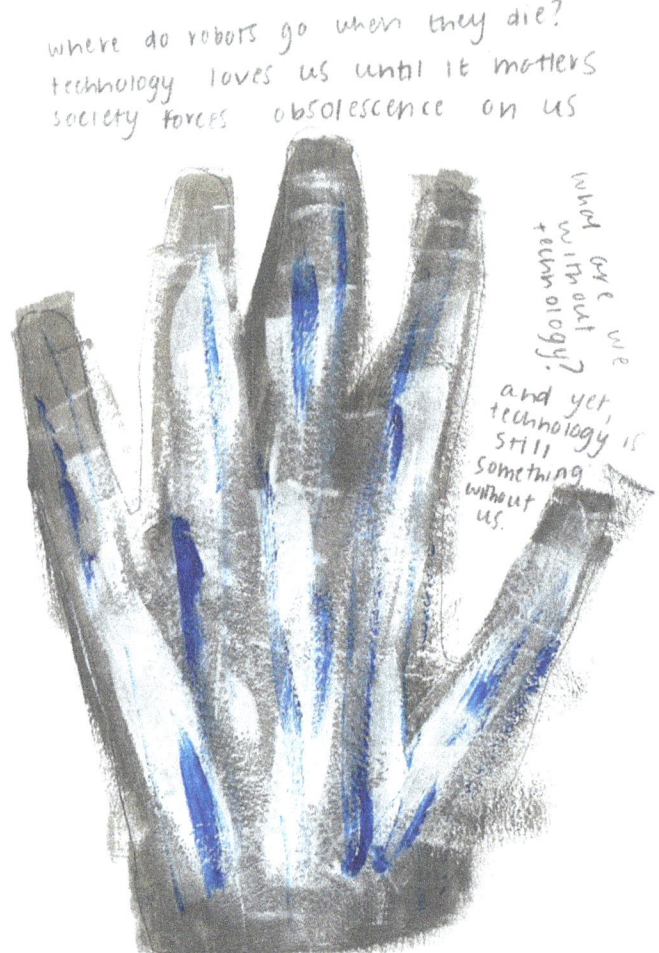

Figure 7.3: Example 2 of students' ecoliteracy hands: "Where do robots go when they die? Technology loves us until it matters. Society forces obsolescence on us. What are we without technology? And yet, technology is still something without us."

While this example of the activity was conducted with undergraduates, students of all ages may use the hand as a metaphorical frame of reference in order to examine and reflect on their relationship with the natural world. For example, in a lesson about natural sciences, students might write words outside of the hand to represent their learning about symbiotic or mutualistic relationships that they are studying. Inside the hand, they might analyze and reflect on the ways human beings function outside of natural systems. Similarly, in a lesson related to behavior in the natural world, students might

Visual Journaling as Ecowriting

Figure 7.4: Example 3 of students' ecoliteracy hands: "How do we live in a world without wanting the next best thing, being slower than the other person, being competitive."

examine the ethics and etiquette of visiting natural spaces and places, using the form of the hand as a visual structure to engage in reflection about human impacts in sensitive ecological contexts.

Ecojigsaw. In this ecowriting activity, students worked collaboratively to synthesize related areas of study using the jigsaw learning strategy. Specifically, students were divided into three groups and each group was assigned a selection of readings related to an ecomedia topic. The topics were: (1) conflict mining and minerals, (2) cloud computing, and (3) electronic waste. The jigsaw strategy is a traditional teaching method that is "helpful in motivating students to accept responsibility for learning something well-enough to teach their peers" (Barkley, 2010, p. 289). After students were prepared as expert groups, they engaged in collaborative visual journaling. A long scroll of paper was unrolled across multiple tables so that students could sit on either side. Students were provided with selected images and graphics related to the topics, as well as copies of the readings, markers, scissors, and glue. Gathering in their groups on either side of the table, students began to discuss and represent their knowledge not only of the issue they were assigned, but also of the interrelationships and effects of the issues. This style of collaborative large scale, or immersive, visual journaling provided students with a generative opportunity to visualize the interconnections between topics, bringing the distant or abstract subject matter into sharper focus. An example is provided in Figure 7.5.

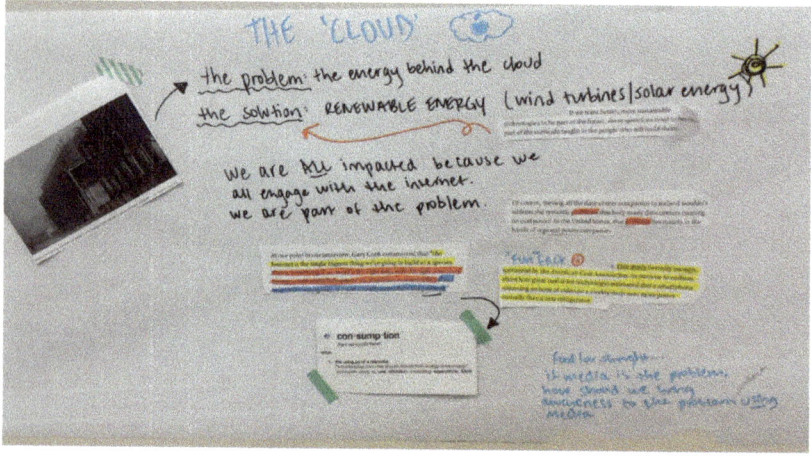

Figure 7.5: Example of a section of the ecojigsaw journaling scroll featuring reflections about cloud computing and problems of energy consumption.

Although the ecojigsaw activity was created within a particular learning context, the general approach may be aptly applied across K-12 contexts. Students of all ages and grade levels would benefit from the social constructivist spirit of this collaborative visual journaling approach, sharing in both the responsibilities of building knowledge using the jigsaw reading method, as well as expressing knowledge and synthesizing ideas in shared creative expression. Many ecological processes lend themselves well to collaborative jigsaw journaling. For instance, elementary-aged students could represent the water cycle as part of their science learning, while older students could represent the physical, social, economic, and political dimensions of a particular topic, such as the production of corn or harvesting of palm oil as part of their social studies curriculum.

Conclusion

Visual journaling as ecowriting may very well prepare students for unknown futures and forthcoming challenges by building in them the capacity for creative thinking and collaborative making. As Knill and Knill (2017) write:

> Of course, we cannot prepare the student completely for the future. We cannot conquer the unpredictable scenes that will arise, nor simulate and role-play the unknown. However, in staying with the arts, we learn how something new is created and comes into our world. This fosters an atmosphere in which the unknown can be encountered more as an opportunity than as a threat. (p. 182)

Similarly, Hagan and Redmond (2019) explain that "[a]rt making and artistic thinking play an important role in cultivating socio-cultural sustainability," a sentiment echoed by many scholars who study and incorporate art to enhance *ecophilia*, or the positive feelings and associations human beings have with natural spaces and surroundings. As shared in this chapter through the stimulating literature related to expressive arts and visual journaling, as well as through our suggestions for interdisciplinary practice and specific examples, ecowriting with visual journaling offers an individual and collective experience to identify, process, problem-find, and problem solve toward a greener and more ecologically focused future.

References

Barkley, E. F. (2010). *Student engagement techniques: A handbook for college faculty*. Jossey-Bass.

Bowers, C. A. (1995). *Educating for an ecologically sustainable culture: Rethinking moral education, creativity, intelligence, and other modern orthodoxies*. State University of New York Press.

Bowers, C. A. (2001). *Educating for eco-justice and community*. University of Georgia Press.

Buzzell, L. (2009). Asking different questions: Therapy for the human animal. In L. Buzzell & C. Chalquist (Eds.), *Ecotherapy: Healing with nature in mind* (pp. 46–54). Sierra Club.

Buzzell, L., & Chalquist, C. (2009). Introduction. In L. Buzzell & C. Chalquist (Eds.), *Ecotherapy: Healing with nature in mind* (pp. 17–21). Sierra Club.

Clinebell, H. (1996). *Ecotherapy: Healing ourselves, healing the Earth*. Haworth Press.

Dewey, J. (1925). *Experience and nature* (2nd ed.). Open Court Pub. Co.
England, L., Carlisle, J., Witter, R., Davidson, D., Holman, L., & Powell, D. (2019). Storying climate change at Appalachian State University. *Practicing Anthropology, 41*(3), 21–26.
Ganim, B., & Fox, S. (1999). *Visual journaling: Going deeper than words.* Quest Books.
Hagan, C., & Redmond, T. (2019). Creative social stewardship, artistic engagement, and the environment. *Journal of Sustainability Education, 20,* 1–19.
Hash, P. (2020a). Articulating literacy: Art making and composition. *English Leadership Quarterly, 43*(1), 19–23.
Hash, P. (2020b). Articulation: Engagement in composition courses through expressive arts. *Journal of Higher Education Theory & Practice, 20*(9), 102–120.
Hash, P. (2021). Expressive arts in virtual spaces: Supporting students during the pandemic. *Art Education, 74*(3), 8–9.
Hash, P., Redmond, T., Adams, T., Luetkemeyer, J., & Davis, J. (2022). Leading with care: Sustaining colleagues and students in times of crisis. *Journal of Curriculum & Pedagogy, 19*(2), 172–184.
Hutchinson, C. (2018). *The simple guide to visual journaling (even if you aren't an artist).* Alacris Publishing.
Knill, M. F., & Knill, P. J. (2017). Aesthetic responsibility in expressive arts: Thoughts on beauty, responsibility, and the new education of expressive arts professionals. In E. G. Levine & S. K. Levine (Eds), *New developments in expressive arts therapy: The play of poiesis,* (pp. 181–193). Jessica Kingsley.
Levine, S. K. (2017). Aesthetic education: Learning through the arts. In E. G. Levine & S. K. Levine (Eds), *New developments in expressive arts therapy: The play of poiesis,* (pp. 181–193). Jessica Kingsley.
López, A. (2014). *Greening media education: Bridging media literacy with green cultural citizenship.* Peter Lang.
López, A. (2020). *Ecomedia literacy: Integrating ecology into media education.* Routledge.
Louv, R. (2008). *Last child in the woods: Saving our children from nature-deficit disorder.* Algonquin Books.
McCaughey, M., & Redmond, T. (2021, June). *In the style of Nora Krug: Visual journaling with the common reading book* [Video]. YouTube. https://youtu.be/a4pZHxYiVOc.
Morrison, A. L. (2009). Embodying sentience. In L. Buzzell & C. Chalquist (Eds.), *Ecotherapy: Healing with nature in mind* (pp. 104–110). Sierra Club.
Orr, D. W. (2009). Foreword. In L. Buzzel & C. Chalquist (Eds.), *Ecotherapy: Healing with nature in mind* (pp. 13–16). Sierra Club.
Redmond, T., Luetkemeyer, J., Davis, J., Hash, P., & Adams, T. (2021). Creating space for care: Sustaining the emotional self in higher education. In I. Ruffin & C. Powell (Eds), *The emotional self at work in higher education* (pp. 120–145). IGI Global.
Shipka, J. (2011). *Toward a composition made whole.* University of Pittsburgh Press.

Shore, C. (2009). The art of healing and the science of art. In K. Luethje (Ed.), *Healing with art and soul: Engaging one's self through art modalities* (pp. 2–13). Cambridge Scholars Publishing.

Sperry, C., & Baker, F. W. (2016). Media literacy. *Social Education, 80*(3), 183–185.

Woodard, C-M. (2020, October). Pantry painting. *SchoolArts Magazine, 120*(2), 36–37.

8. Disrupting Hierarchies of Power and Uplifting Environmental Justice through Collaborative Ecowriting

ALEJANDRO OJEDA, ELMER ORTEGA, JENIFER RAMOS, VANESSA ROMERO, NEIDA SANDOVAL-LOPEZ, BENJAMIN THOMPSON, MARÍA VERÓNICA VALERIANO, AND JAZMINE VEGA LOPEZ

Introduction

As a group of undergraduate students at UCLA, we met in the Critical Digital Media Literacies (CDML) class taught by Professor Jeff Share and Teaching Assistant, Andrea Gambino. All of us were enrolled in different undergrad programs, such as Public Affairs, Education, English, Film, Political Science, History, Sociology, Gender Studies, and Psychology. We were brought together by the wide scope of Critical Media Literacy during the middle of the COVID-19 pandemic. About a dozen of us had taken an environmental justice class the previous quarter with Jeff and at the start of this quarter, we were given an opportunity to take over some of the teaching in this virtual classroom to integrate the skills and concepts we had been learning in both classes. This was mind blowing! An educator, the person we were always taught was the ever-powerful Oz, was handing over the reins. We are the eight students who grabbed those reins and taught ecowriting in our Critical Media Literacy class. In this essay, we want to share our experiences as facilitators of this collaborative process.

From the beginning, we knew we were disrupting the traditional classroom routine. During our first meeting as a group, we started with a raw idea of the topics we wanted to address. As the quarter progressed, those ideas energized the process of collaboration, hard work, and genuine interest. Unlike most classroom lessons, this was not a requirement and we were not getting graded for it, yet it became one of the activities we have been most proud of in our undergraduate careers.

Four weeks into the fully online quarter, we began meeting on Zoom, outside of class every Wednesday afternoon with Jeff and Andrea, to organize the lesson we would lead with our fellow classmates. We knew early on that we wanted to focus on ecowriting and embrace a unique, peer-led structure. As a group, we decided to design an activity where we would create collaborative poems based on students' photographs, thereby featuring everyone's voice and creativity. One week before the poetry writing, the class voted on two topics they wanted to explore related to the environment: environmental racism and biophilia (the love of nature). For the next class, we asked all the students to photograph the impact of environmental racism in their community, as well as take or find pictures about their positive relationships with the natural world, their biophilia.

When we began the poetry writing in class, Kathy Lizaola, a former student from the previous quarter's environmental justice class, shared her ecowriting with the whole group (see Kathy's poem, *A Meaningful Purpose*, on pp. 121–122). We then separated into four breakout rooms with two facilitators and six to seven students in each room. In two of the breakout rooms we discussed biophilia, and the other rooms focused on environmental racism. We asked the students to share the photographs they took and discuss what their images represented to them. This brainstorming session enthusiastically generated a bunch of ideas for our poems.

We had the freedom to compose our poems as we chose, including anything we wanted, deciding how long or short it would be, and what type of poem we wanted to create (although all of them ended up being free-verse). When each poem was finished, we practiced reading it out loud together, then recorded each person reciting the line they had contributed. After everyone had an opportunity to reflect and share their pictures and writing in the breakout rooms, two facilitators (María and Elmer) took the images, writing, and voice recordings from each group and edited them together. We culminated with multimedia examples of ecowriting; four digital video poems narrated and photographed by students addressing biophilia and environmental racism.

Throughout the many years that we have spent as students, we have learned that education usually follows a typical structure, one in which students sit in a classroom, memorize information, and follow a rigid routine of note taking and test taking. This activity gave us the autonomy to learn about a topic in a way we wanted to and be guided by our own peers, aiding in a cycle of learning and teaching.

We initially felt intimidated by the lack of faces on the screen, as only a few students had their cameras on. In one of the breakout rooms, Neida recited the ecowriting poem that she created during her previous environmental justice course, while other facilitators showed the photographs they had taken (see Neida's poem, *On Vacation*, on p. 127). By sharing our own work in this way, we demonstrated how pictures can inspire words, and words can be used to create pictures. This served as a foundation for understanding what we were asking from the group and helped to establish a level of trust and comfort between facilitators and students. When we started the activity and had the students share their photographs, we realized that the learning environment didn't feel like the one we were accustomed to. Instead, we were all collaborators and allies in our shared educational space. Although the task of the group may have been to create a poem in tandem with the images, our main priority was ensuring that we could learn from each other and the different perspectives we shared.

Part of our success with this experience may have been due to the dynamic of the lesson itself. There was not a rubric against which contributions would be measured. Rubrics often cause students to feel anxiety about the quality of their offerings (e.g., *Is this right? Do I sound stupid? Is this even coherent?*). There are countless assessments that students endure for the sake of measuring "student success," but sometimes what they actually measure is student inequality and the lack of resources and support students receive. Instead of using rubrics or tests, we encouraged peer feedback. As students gave feedback to each other, their contributions and affirmations were more meaningful than a subjective score or rating that often stifles creativity.

As facilitators, we were extremely excited about this project and our enthusiasm was contagious. While students were sharing their work, we responded with our genuine positive reactions. This motivated other students to send affirmations in the Zoom chat regarding their peers' work. By providing the opportunity and encouragement for students to offer positive feedback to their peers, students felt more comfortable sharing their creative process and expressing their appreciation. Throughout the activity, we thanked the students for sharing and asked probing questions that encouraged reflection and participation.

The importance of activities such as this signifies a balanced and evocative rhythm, one where teachers work in tandem with the needs and voices of their students. By promoting activities that recognize students' wholeness, our assets and not our assumed deficits, the goal becomes engaging with one another and understanding the big ideas. We can't expect one standardized test or one style of teaching to work for all populations of students.

Our Pedagogical Process

When planning for this project, we prioritized interaction and collaboration so that students would maintain engagement and interest. If students encountered difficulties while creating the poem, we were prepared to support them. As students in the Department of Education, we have learned about various strategies to scaffold active participation, discussion, and reflection to deepen critical thinking, nurture empathy, and enhance relatability. One of the primary ideas that grounded our decisions was our commitment to the process being more important than the product, thereby ensuring that the focus was on learning rather than just completing an assignment.

In order to focus on the process, we used guiding questions for the students to explain the stories behind their pictures. The questions included: *Why did you choose it? Where was the picture taken? What do you want us to know about it?* This allowed for the rest of the class to put themselves in the speaker's shoes when listening to the context connected to each picture. Hearing each other's personal narratives and background stories was crucial for creating the poems together. This also gave us an opportunity to gain trust with one another and create a tight-knit community, something that most of us had not experienced before.

Throughout our activity, we wanted to create a welcoming space for students to participate, so we began by explaining our plans and expectations, checking for understanding, and providing opportunities to adapt the structure to their needs. As students, we have found it helpful to have time to prepare our responses without the pressure of being put on the spot and required to share. A strategy that helped us keep the lesson moving and lower student anxiety was to ask students to share in the same order they had done before, always with the option to pass, so there were no surprises.

Timing was also an essential part of the lesson and it was easy for us to lose track of time while engaging in the conversations that involved personal sharing. Since our goal was focused more on the process than the product, we wanted to listen and learn from each other more than keep to a set time schedule; this balance was difficult. Providing students an open space to share and discuss their personal relationships with broader environmental issues was especially enlightening, allowing a deeper appreciation of the work that they were completing.

Reflecting on the Power of Positionality

We all come from different backgrounds and cultures, and research suggests that student engagement increases when we have teachers who look like us and can relate to many of the stories, barriers, and obstacles that we share. Through this activity, we learned that student voices are necessary in educational environments. Since many of us are first-generation college students, we recognize that it takes much more than just student drive and tenacity for success; institutional support is fundamental. The voices of students must be recognized and included in the classroom, and any excuse not to do so is accepting and encouraging a failing system.

Jenifer reflects on her experiences in school, writing:

> I often think about the 'good' teachers and the 'bad' teachers I have had. The good teachers always cared about me, about how I was doing in school AND outside of school. The bad teachers barely taught, yelled at students frequently, and focused more on anything besides educating us. Student voice is necessary in an educational environment. I only started discovering my voice in high school – but not in a class, rather a club. As a member of the speech and debate club, I learned to recognize the strength and power of my voice as a student, and I learned to love it. I began to hold my teachers accountable for their lack of interest in properly educating me. After I graduated high school, there were countless times when I was met with disengagement and ignorance by university staff. In only a few classes did I have professors who uplifted student voices and cultivated student teacher partnerships. A classroom is a shared space, so why is the structure of education often silencing and unequal?

Reflecting on this space and her previous experiences sharing in classes, Jazmine writes,

> As a student who is hesitant when it comes to participation because of anxiety, this activity was way out of my comfort zone. I came in a bit skeptical about this opportunity and feared the worst-case scenarios possible, such as *'what if the students don't like the assignment, what if we run out of time, what if no one participates?'* However, working in groups allowed us to create a community in a span of a few minutes. Some of us were in classes together before but never got to really hear about each others' experiences and ideas. With the help of the photos we were asked to bring in, we were able to learn about our communities and the places we called home. We were also given the space to work as a group, and communicate, plan together, share ideas, and manage our own time. I felt a safe and comfortable environment that was created for each of us to contribute to a project that was outside of my comfort zone, and I was able to overcome my fear of social interactions and public speaking. A lot of shells were broken.

Conclusion

Traditional classroom settings teach students to cram information, which they often forget soon after class, instead of engaging students with the material and creating a joyful learning process that contributes to positive feelings and information retention. When a teacher simply reads a plethora of slides and expects students to memorize the given information, it can be difficult to find pleasure in the learning process. The brain is more likely to retain information when the process of learning is enjoyable and meaningful.

Throughout the careful planning of our lesson, we wanted to share what we had learned with our peers in a way that would embody what we often forget education can be: the engagement of ideas through collaboration and joy. By using the arts, such as photographs and poetry to allow students to collaboratively express their passions, creativity, and understandings of the environment we were sparking joy and making learning memorable. Although the end result of our ecowriting was beautiful, the process was even more so because of the high level of enjoyment and engagement from every student.

After the activity came to a close, our classmates vocalized their appreciation of being led by peers, specifically noting the learning that had taken place in their smaller discussion rooms. This is where we saw a connection between environmentalism and critical media literacy. In giving students the opportunity to view environmental media from the position of the creator rather than just the consumer, students and facilitators alike were motivated to reflect deeper and approach all media creations from a new perspective.

The learning continues for us facilitators, as we have not stopped meeting weekly, even months after the class ended and most of us have graduated. We are writing this essay in the hope of inspiring others to incorporate ecowriting in their own classrooms. By prioritizing the process over the product and integrating the arts, we can empower students as partners and collaborators in a space dedicated to learning and teaching.

9. Grand Appreciation for all Things Natural

ROSE WHITE

I've always loved flowers, possibly because of my name … but also because of the many hours I spent, as a child, exploring my grandmother's lush and varied gardens – expanses that covered her entire front and back yards.

More than two decades ago – and while I was still a classroom teacher, the Los Angeles Audubon chapter in Culver City gifted me with a week's birding experience in Greenwich, Connecticut – all expenses paid. The prior fall, I had taken my inner-city kindergarten class to their field trip site in a lagoon near the Ballona Wetlands in Playa del Rey, prepped them well-replete with the knowledge of how to use a microscope to investigate the multitude of organisms that inhabit such waters. In fact, when they found out that my students were so young, they negated the use of microscopes on that trip to the point that both the students and I protested in disappointment. That was an eye-opening and wondrous experience for all involved, and the next thing I knew, one of the directors of student engagement had me on the phone interviewing me for that marvelous summer experience back East.

While in Connecticut, I had an opportunity to walk old growth forests, examine rock walls, go birding in the daylight and at night, explore the shores for horseshoe crabs, dine in an ancient barn that had been reconstructed on the site (where I also learned some spectacular new recipes), take classes in ecology/conservation/recycling-reuse/activism and much, much more. I still have two t-shirts that I purchased on that trip, made by the camp counselors – one that was hand-illustrated with acrylic bird droppings that I wear every Halloween, and another that shows an array of mushrooms that we came across in the forests during that week. I get a number of compliments about that mushroom t-shirt every time I wear it.

Undoubtedly, my teaching of science, and all things natural the following year was enhanced immensely as I turned my school environment into a mini-Audubon site. We went birding for Plovers, Killdeer, pigeons and gulls; one even built a nest in the middle of the faculty parking lot and had to be protected by the school's custodian. We adopted several trees on the grounds and chronicled their changes over the months through close observation/discussion and art. We collected seed pods that fell from the trees and used them as templates for reverse tie-dying swatches of cloth (THAT was the craze way back then) that we quilted and turned into TV Watching Pillows for Father's Day gifts.

My own regard for nature went through a paradigm shift that fateful summer as well, and has been evolving ever since. I have an incredible regard for the natural surroundings of my own personal urban landscape. I "built" a few "rock walls" on my property after I had some retaining walls refurbished and elected not to throw the pieces of the old structures in the trash. Of course, you can barely see them now, as they have been overgrown with ivy. I rejected the notion to install solar panels on my roof because the companies that gave me bids said that I would have to cut down or severely cut back the two 70-year-old Italian Stone Pines and the Canary Island Pine that flank my house and cover my roof. Ironically, those three trees have kept my house on the "cool side" the 40 years that I have lived here and preserved the roof that I only recently had rebuilt.

I don't consider myself a formally trained nor self-taught birder … just a curious observer who consults the multitude of books on birds I bought when I was an unretired teacher. I have spotted on my property owls, hawks, woodpeckers, hummingbirds, mourning doves, blue jays, and some tiny noisy birds that look like gray "radishes" with beaks. I even saw a brilliantly colorful orangey-yellow bird the other day flitting around a neighbor's property; can't decide if it's a type of Oriole or a Tanager. I've potted milkweed throughout my garage-top succulent garden to attract migrating Monarchs (the kind I saw once when I visited Michoacán, Mexico), as well as other entertaining pollinators.

During the onset of the Covid pandemic in 2020, I had the privilege of "caretaking" several families of Black Phoebes that decided to use my backyard as a nursery and flight training pad for their various offspring. What a delightful distraction that was for me at such an anxiety-producing period in our nation's history. I kept a diary from May 'til September of their exploits, which brought me great joy and validated the fact that, when most needed, nature can be a tremendous source of peace and solace in one's life.

Part III: Student Ecowriting

10. A Meaningful Purpose

KATHY LIZAOLA

There are many things we take for granted
Media constantly keeping us distracted
Diverting us from actually reflecting on what truly matters
The present
Going out on walks in nature and seeing the essence of what true and pure beauty really is
Birds chirping, flowers blooming, sun shining, and rain pouring
Very comforting
Beauty is everywhere
Sometimes we don't take the time to stop and stare
Others do see it, but have different perceptions
Some see nature as something to be exploited and commodified
Others see it as something to be nurtured and respected
I capture beauty in my drawings
Drawings of people, flowers, trees, insects, and animals
Anything that signifies beauty in my eyes
But as I reflect on my appreciation for nature, I begin feeling more like a walking contradiction
Using so much paper, pencils, crayons, and paintbrushes
All of which come from trees
Trees that come from nature
Trees that provide us with oxygen, shade, and life
Feeling guilty for reinforcing this cycle of over-consumption and exploitation
But millions of people are engaging in this cycle, too
Globally
Collective action is the only way to fight against these corporations harming our environment
Collective action begins with individual choice
Choosing to make a difference or remain silent
Questioning what I can do as an individual

My passion for art can be used as a way to expose
Expose these corrupt corporations
Bring awareness to environmental racism
Rebuke arguments saying "climate change is a hoax"
Highlight the importance of believing in Science
Encourage others to engage in environment-based organizations
Inspire others to be out in nature
Motivate others to use and not waste everything they take from nature
Motivate others to explore and appreciate the things we take for granted
Rather than feeling guilty, I can turn my actions into something positive
Using art to inform people about the Truth
Giving back to nature
Being an advocate and a leader
Having a mutual understanding of what's really happening is crucial
Everyone has a meaningful purpose in this world
Use it, don't waste it

11. The Life of a Fast Fashion Garment Worker

GABRIELA VENEGAS

At dawn, I change into my old clothing
There is nothing that I am more loathing

Walking to work, the streams look depleted
Showing how our ecosystem is extremely mistreated

Hazy orange, gray clouds fill the skies
Just to produce your beloved Levis

Passing mounts of garments that have been tattered
Symbols that the dignity of workers have been shattered

I enter the immense, cold garment factory
Where my work will never be satisfactory

I cut the fabric that will make "yesterday's" clothes
Which will take decades to decompose

I rapidly sew to support my family
Being fairly paid is my fantasy

I get paid an unsustainable, low wage
Just for your clothes to be all the rage

Every day I create the newest, hottest trend
For people to over shop and overspend

Workers are in constant misery
For the sake of the fashion industry

I desire there would be more legislature
To protect me and mother nature

12. *Is It Even Air?*

GISELLE VILLANUEVA

It feels like poison is in the air.
How am I supposed to breathe with no air?
I step outside and I breathe polluted air.
This cannot be fair.

Could it be the freeways surrounding my community?
The lack of green spaces, the lack of opportunity?
Diesel engines, cars, buses, trains, industries, deforestation are a bigger threat every day.
And it just doesn't go away.

My community is hurting.
Lynwood.
Screaming for help.
Worried about their health.

The problem is big corporations are aware but they pretend it's not there.
It is so maddening ...
That these things are still happening.

Battling the need for trees, many trees.
But how can we move forward?
Don't care about the government, they take it all down anyways.
We can try many ways.
Corruption and Capitalism are stronger every day.

I see so much pollution, but where is the solution?
I wish for it all to disappear.
But those wishes seem so unreal.
The U.S. takes and takes but never gives.

How much longer can Mother Earth take this?
Is it days, months, years until she can no longer resist?
We cannot dismiss

The need for action.
Don't want to leave this world with dissatisfaction.
When my lungs are no longer strong enough.

I fear for the upcoming generations.
To live a life with many health complications.

Despite it all, I still have hope, hope the world will turn.
Hope for clean air and water for all.
Hope for the sea animals.
Hope for crops.
Hope for forests.
Hope for aquatic ecosystems.
Hope for healthy days.
HOPE for good quality air for all.

This can be possible.
Reducing vehicle exhaust fumes,
Reducing exhaust from industrial plants,
Exhaust from factories,
Reducing fossil fuels,
Increasing green spaces.
Spreading awareness of climate issues

We need to prioritize the environment in low-income communities.
Share resources to beautify local environments.
Together we can build a better world for the environment.

13. *On Vacation*

NEIDA SANDOVAL-LOPEZ

I put my phone down for a second to look around the place
And as I scoped, I couldn't help but notice an abundance of similar traits
Everyone was smiling
Recording with their phones held up in front of their face
Trying to scoot the furthest back to perfectly capture the whole space
We all know that this experience could never be replaced
Savor the moment, we advise, because we might not ever see it again
How many more years till we come back? Eight, nine, maybe even ten.
But isn't it weird to think that it wasn't like this back then?

It all started when a man perceived the planet as profit
He said, as long as we privatize our land, we could take advantage of it
We slap the name 'resort' on it and have people drive hundreds of miles
Spend thousands of dollars to hang glide over it, swim in it, find peace in it and call it a vacation
But these are just the results of a nation that was established through colonization
They've destroyed you and built over you without any hesitation
Now just to see you, we even have to pay
Why did we choose this life?
When we could have had you for free every day

I feel your pain
We exploited you and killed you for our own personal gain
Violently reduced the number of years for your life to sustain
And after realizing this, I have grown more concerned
How did I go all these years uninformed?
I cannot sit back and watch
I will use my words to share what I've learned
Maybe then we can live long enough to see this world transformed

14. Hard to Appreciate

KEIMORA NETTLES

How can I appreciate the natural world, when the one I live in was designed to treat us unnaturally?

My people. My family. My community.

I have been naturally unable to appreciate my environment.

Compton. Watts. Inglewood. Crenshaw.

These cities are nothing like,

Westwood. Santa Monica. Bel Air. Beverly Hills.

I learned that outside was a place where people are targeted.

So it's hard to appreciate.

"Get inside before the street lights cut on!"

"The police drove around there earlier today."

"Don't hang out at the park!"

"Get in this house!"
It was all too familiar.
So I stayed inside.

Reminded of the horrors of being Black, living in LA, and the constant task of checking your surroundings.

I did not have time to appreciate the trees, leaves, ocean, or outdoor spaces.

Not because I did not want to, but because nothing has ever been accessible to us.

Or was it because the bloods and crips roamed around ready to interrogate anyone who looked "out of place?"

Or was it because everything was "out of the hood" and it would be a 45-minute drive to enjoy the things that my community could not?

There are no farmer's markets. But there are liquor stores.

There are no garden spaces, but there are farmers markets and healthy food options 20 miles away.

It's hard to appreciate.

Especially something I never knew I was lacking until I came into spaces where people had it all.

A deeper understanding of the natural world.

The same world I also live in, but had no reason to believe I rightfully had access to.

Why is it like this?

Why is it hard to have access to the things most people have?

Yet, here I am, still today looking for the answers but reminded of the pain as I look at the LA Riots, racial divide, and tensions between communities.

This is where I begin to see the unfortunate aspects of the natural world I know.

The spaces where it is not safe to enter, drive through, or live in because anything is bound to happen.

So yeah, it's hard to appreciate.

15. Feels Like Home

ANTONIA BURGARD, LUDMILLA SEMSKOW, LEA IRINA HEUING, JANA KRANZ, AND PHILIPPA WITZENHAUSEN

This is an example of digital ecowriting. This video poem was created by a group of five pre-service teachers in a masters degree teaching credential program who completed a course facilitated by Dr. Frauke Matz at Westfälische Wilhelms-Universität Münster. The students first explored the Botanical Garden in their city by themselves, focusing on their experiences of seeing, hearing, smelling, feeling, and engaging with their natural surroundings. After 10 minutes of independent exploration, they worked together in groups of five students and shared the words they chose to create a group poem. Finally, they collaborated to create a video poem with their cell phone.

View "Feels Like Home" via this URL or the QR code: https://tinyurl.com/mu2a2byr

Figure 15.1: Still image from the video poem "Feels Like Home."

Figure 15.2: QR code for digital access to the video poem "Feels Like Home."

16. Divisions That Destroy Us

SARA FERNANDEZ

I look at the scar. We call it the scar because it is such a deep wound that has yet to fully heal. The scar is tall, pierces the earth, and creates a division. A division that is so high not even I can see past it. I don't remember what the other side looks like, will I ever be able to see it again? Every day, I prance around the Arizona desert. Trying to find food, trying to find water. But the big towering barrier's shadow looks at me, mocks me. It has stolen these things from me. Men with their big trucks and machinery deepen the wound every day. And they take my water, my food, and my shelter. And yet the ginormous scar wins again. Sometimes on my daily runs to fetch food I see fellow bobcats and wildlife fall to the ground, lifeless. Their bodies lying dead and the barrier just looking, glaring at them.

But I know there is hope. I see humans who fight for my life. Humans who mourn the creation of the wound and who want to let it heal. I see the big men with the big machinery destroy our lakes where we get fresh water from. I also see them blow up their sacred sites. I see these people, angry at the barrier, angry at the scar it has created on the land and on our lives.

These people want to see me survive. For my existence to them is crucial to theirs and that of their ancestors. They realize that the land is in pain and needs to heal, and with this healing it means we get to live one more day.

Recommended Reading

Gilbert, S. (2020, October 31). 'An incredible scar': The harsh toll of Trump's 400-mile wall through national parks. *The Guardian*. https://www.theguardian.com/environment/2020/oct/31/trump-border-wall-wilderness-wildlife-impact

Jordahl, L. (2020, November 1). A year of devastation in Arizona's wild lands. *New York Times.* https://www.nytimes.com/2020/11/01/opinion/trump-wall-arizona-environment.html

Parker, L. (2019, January 10). *6 ways the border wall could disrupt the environment.* National Geographic. https://www.nationalgeographic.com/environment/2019/01/how-trump-us-mexico-border-wall-could-impact-environment-wildlife-water/

17. *Water*

RUCHA DESHPANDE

The sky arches above me as I dangle my feet off the side of a boat. People are chatting, yelling, laughing. The air whistles right by my ear. Seagulls squawk as they pass by overhead. A distant horn blows. The waves splash gently beneath me.

I look up at a distant blue expanse too far for me to reach. I look down and take the plunge.

This is a different kind of blue, I realize. There are no voices. Instead, I am enveloped by a vast, sapphire silence. Bright flashes of color enter my vision as I watch vibrant fish pass by. Light filters through a neighboring kelp forest, stalks undulating lazily with invisible undercurrents.

I soak it all in. I am immersed in the sea. I am filled with the same sense of calm that I feel when watching waves breaking on the shore, surrounded by the colors of the setting sun.

But what if this vastness wasn't as silent as I thought? If the ocean could speak, what would it say to me? Would it talk, plead, protest? Or would I be greeted by a more choked, coerced silence? Would I perhaps feel something other than reverence or peace?

The ocean is at once beautiful and haunting, at once comfortable and mysterious, at once silent and deafening. With no Earth below me nor sky above, I remain suspended beneath the waves, listening.

18. *Elena and the Mountain*

Vanessa Romero

The sky is still dark when Elena reaches the trailhead, she looks around and it appears she is all alone. The moonlight barely offers a glimpse of the outline of the forest. There is a crisp, coldness in the thinning mountain air. With each inhale she takes, her breathing becomes faster and shallower, there is a slight burning sensation in the back of her throat. It takes some time for Elena to get used to the change in altitude but no sooner as she begins her ascent, and her body warms, her breath starts to find a rhythm.

About a quarter mile into the hike a gust of wind blows the trees and they begin to sway back and forth. Elena notices a figure standing in the distance, another lone hiker on the trail. She approaches the hiker and offers a friendly "good morning" to her, they strike up a conversation and the lone hiker introduces herself as Dhara. Elena decides it would be nice to have a companion for the rest of the hike and Dhara agrees. The sun is starting to rise in the distance, there is rustling all around in the bushes and from high above in the trees they can hear the birds chirping. The forest is starting to wake up for the day.

Elena and Dhara begin at a steady pace, about a 45-minute mile. The first few miles up to the saddle is nothing but incline and Elena's legs are burning. She begins to breathe heavily again and the wind chill feels like pine needles on her face. Elena has never hiked this trail before, and she is struggling with the elevation gain. Dhara notices and comes up with a plan, she tells Elena that at each mile they will stop and take a 5-minute break. Elena likes the idea and so the two hikers continue their ascent. Even as Elena's pace begins to slow Dhara stays with her, encouraging her to make it to the next mile marker and then onto the saddle.

The sun starts to peek out above the tree line and Elena can now feel the warmth of the rays on her face. The air begins to warm and all around the squirrels are scurrying about the trees and the lizards are warming their cold-blooded bodies on top of the rocks, paying no mind to Elena and Dhara. There are paw prints and scat along the trail and according to Dhara they are from a black bear. Dhara explains black bears are very common to the area around this time of year and the best thing to do is to keep moving and be aware of their surroundings. Dhara has been out on the trail many times before and shares her knowledge of the area with Elena. There is so much to learn about the local wildlife, the flowers, plants and trees of the forest and Elena is so intrigued she soaks up all of the information like a sponge. Dhara also points out along the trail where they can find sources of fresh water to refill their packs. The time flies and Elena no longer can feel the burning in her legs and her breathing has steadied. Before she realizes it, they have reached the saddle where they stop to have a snack, take a breather and chat about making it this far; a sense of accomplishment is starting to come over her.

Then, before long it is time to continue on their journey up the mountain. After the saddle they reach a section of the trail that feels almost flat, a nice reprieve for Elena. The trees clear away almost in anticipation of the two hikers' arrival, the sun is shining bright now and all around the bushes are covered in these beautiful blooming flowers. It is like the forest is welcoming Elena and Dhara and the scene before them is their reward for making it this far. Elena can feel the energy all around her and soaks it all in, who needs coffee when you have the forest to invigorate you. Elena knows the next few miles are going to be the toughest, the push to the summit always has the steepest incline and they will encounter a very quick elevation gain. She takes in the splendor around her and then proceeds into the trees again.

Elena and Dhara notice the trees start to become sparser, and, in the distance, they can see what appears to be the summit, but it is still a ways away. The trail starts to narrow, it becomes very rocky and steep. Elena knows they are almost there, but she is tired, her arms feel numb, her legs are burning again, and her pace has slowed. Dhara pushes Elena and tells her to find a rock or a tree along the trail, some marker, and set a goal just to reach that marker. Once Elena reaches her goal, she finds another marker ahead of her and then reaches that spot, takes a break and sets another goal. Elena proceeds like that the rest of the way up the mountain and the summit no longer seems so far away. The difficult summit before her has now become more manageable and it is not a race but a journey. Elena finds her pace and before she knows it, she can see the very top, it is within her reach. She can see the fluffy clouds that blanket the entire sky. Elena can feel a sudden burst of energy and excitement run through her body; she has reached the summit. All of the pain, the discomfort, the doubt, the struggle is all worth it when she reaches the top. She turns to thank and rejoice with Dhara but somewhere along the way Dhara disappears. Elena can feel her presence though, and smiles, she leaves everything behind her and lets the feelings of happiness, serenity and accomplishment overcome her.

19. *The Evolution of the Apple*

ELMER ORTEGA

Capitalism has brainwashed people into believing that making money at the cost of the environment is acceptable. I was raised by parents who come from a ranch in Mexico, located on a green mountain top, filled with natural life. Since I was small, my mom in particular, has always preached the importance of not killing or destroying because she believes that every life or aspect of nature should be preserved, even something that may appear insignificant to most, like a single ant or unappealing plant.

I ultimately inherited this perspective and appreciation for all natural elements from my mother. While my mother herself was not completely tied to her Indigenous identity like her mother, my grandmother's Indigenous values of preservation and care for natural life resonated with her. She adopted these aspects of her cultural identity and then passed them on to me. I recognize the importance of these values; therefore, I try my best to uphold them. But, I fail and feel guilty because everything I do and everything I buy contributes to a capitalist American culture that undermines what I have been taught about nature.

In my mother's ranch, which is relatively remote, an apple is cherished since it is seen as nature giving back for taking care of its tree. However, in the United States, the word "apple" is associated with a tech company, which is responsible for contributing to a large carbon footprint. This company takes away meaning from the true beauty that is a gift from nature, *the apple*.

Apple strategically creates new products annually with slight improvements to take advantage of the brain's attraction for novelty. People talk more about the craze and hype around a new Apple product rather than the millions of metric tons of carbon they emit into the atmosphere.

By killing nature we are killing ourselves. Was the company named Apple as a way to rebrand a part of nature? To brainwash people into forgetting about the nature they killed? Could it be a scheme like the Catholic church that rebranded Pan the Greek's god of nature into looking like the devil in order to deem it acceptable to exploit our nature?

Recommended Reading

Vachon, P. (2018, October 13). *Forbidden fruit: An exploration of the apple in historical and popular culture.* Chowhound. https://apple.news/azNUqfJ7_TAaYe !eC-d6BuQ

20. Gardening in the Time of Covid

ESMERALDA OROZCO SANCHEZ

This is the small pot that started it all. While the whole world was in a panic and each day felt the same, this little pot brought me joy once again, at least for a little while. It came with a packet of sunflower seeds and a little brown circle that when wet with a few drops of water expanded into a sufficient amount of soil. Seeing this process felt like magic and it excited me for what was to come next. I followed the instructions for days, which explained that the soil had to be sprayed with water as to not over water the seeds. To my great disappointment, the seeds did not blossom in the small pot. To this day I still do not understand why they didn't blossom given that I had followed the instructions word for word.

Fortunately, my mom had a bigger pot full of soil in the front yard that she had recently bought, and she suggested we plant them there instead. They blossomed within two days and grew to their full height in a few weeks. Their beauty convinced me to join my mom in adding to her small garden and during those first weeks of quarantine we were able to transform a nice sized portion of our front yard into a slightly bigger garden. Since the soil in Pacoima is very dry, making it hard to grow anything in it, my dad helped us by building a bricked space which we filled with commercial soil. At first, we filled the area up with different types of flowers that both my mom and I had chosen. My mom later planted chilies, tomatoes, and potatoes. The tomato plant died, but the chili plant grew beautifully, and the potatoes came out small but delicious.

Starting this garden with my mom during such a rough time was very helpful for both of us. It not only helped us take our mind off of the pandemic, but it also helped us form a stronger bond. About three weeks ago, my mom suffered a great stomach pain which landed her in the hospital for a week. She underwent two surgeries and during this time I was under a lot of stress which caused me to neglect the garden. It wasn't until after my mom's third night in the hospital that I finally remembered the garden and realized no one had watered the plants in days. It was already 10 p.m. when I came to this realization and although I had no motivation to get up and water the plants at such a late hour, I began to think about how sad my mom would be when she came back home to a dead garden. After that, thinking about how special this garden was to my mom and I, gave me the much-needed motivation to continue caring for it. Now that my mom is home, she is in love with her newly blossomed *Noche Buenas* [Poinsettias], which had been a simple green plant for months but now that December has approached, they have begun to bloom. This garden has become a source of happiness for my mom and me and I have come to realize that it is also a stress reliever, one that has been extremely helpful in these stressful times. As a result, I plan to continue gardening and adding more plants and flowers to our garden once my mom is healthy enough to help me plant them.

21. *The Manifestation of Taoism in Environmental Protection*

(Alice) Yanan Sun

One day Lao Tzu led his students to a vacant field and invited an old farmer to explain to his students how farmers farm. The farmer explained that there are eight steps in farming: plowing the ground, watering the ground, plowing the ground, leveling the ground, drying the ground, spreading fertilizer, sowing seeds, and leveling the ground. Lao Tzu asked the old farmer to demonstrate the procedures in farming, and then Lao Tzu took his students and began to plow the soil together. Lao Tzu and his students repeated after the old farmer and plowed the ground, watered and leveled the ground, spread fertilizer, and sowed the seeds. After a tiring day, Lao Tzu took the students back to the Taoist monastery. The students fell to the ground one after another and did not understand why Lao Tzu did not preach but took them to work in the field.

The following dialogue ensued:

STUDENT: Teacher, why did you take us to work in the field instead of preaching today?
LAO TZU: I have been preaching all day today.
STUDENT: Teacher, can you enlighten me with what you mean by preaching all day today?
LAO TZU: What I teach you today is environmental justice.
STUDENT: Teacher, what is environmental justice?
LAO TZU: Environmental justice can be summarized in three idioms: Nature and humans are one [天人合一], three talents steal each other [三才相盗], and all things are in harmony [齐同万物].
STUDENT: Teacher, can you explain the meaning of these three idioms?
LAO TZU: Nature and humans are one: humans are part of nature. However, people have lost their original nature and become inconsistent with nature

because they have formulated various laws, regulations, and moral codes. The purpose of human practice is to return to true self, break down these barriers imposed on people, liberate humanity, return to nature, and achieve a spiritual realm of "all things and I are one." Three talents steal from each other: humans have a causal relationship with all things in nature. The so-called good is rewarded for good, and evil is rewarded for evil, which also exists between humans and nature. Nature is not just a passive receiver, and it is not predatory by everyone. It may also retaliate against humankind. Therefore, one cannot claim to be a conqueror or consider himself or herself the most precious. To respect nature is to respect humankind itself. Harmony with all things: Although there are differences between the species, between people, or between different cultures, different nationalities, and different beliefs, these differences are always integrated and harmoniously balanced. In the view of Taoism, anything that separates from the system will suffer bad luck. Therefore, we should pursue harmony in nature.

STUDENT: Teacher, I understand now. But why are we farming today?

LAO TZU: That is because I want to implement place-based pedagogy in preaching to cultivate your love of nature. Place-based pedagogy allows the student to learn to take care of nature by understanding where they live and taking action within their power.

STUDENT: Thank you, Teacher!

The ecological view of Taoism emphasizes that humans and nature are sharing the same fate so that humans must coexist and move together with nature. Taoism also emphasizes the balance between heaven, earth, and humans. Taoism reminds people to care for nature.

22. *Armando*

María Verónica Valeriano

INT. ARMANDO'S BEDROOM – DAY
A young boy, ARMANDO (12), wears a large hoodie and lays on his bedroom floor. His big, round, black eyes fill with concern as he hears the power plant's SIRENS go off.
  He gets up and runs into his mother's room.
                    ARMANDO
                   (In SPANISH)
          Ma! Cover your nose! Hold your
                    breath!
He jumps on her bed and places his small, chubby hands on his mother's face, covering her mouth and nose.
  IRMA (early 50s), lays on her bed covered in a thick, mustard yellow blanket. Her eyes are tired. The wrinkles on her face noticeable. Her nightstand is full of amber prescription bottles and a plastic water bottle.

INT. CLASSROOM – DAY
ARMANDO walks into a Science classroom with a backpack twice his size. Student work fills the classroom walls. Students enter the classroom, their voices filling the room.
  A curly-haired woman, MS. SANDOVAL (early 30s), gathers graded homework assignments from her desk.
                    MS. SANDOVAL
          Quiet down, everyone! Take a seat.
She starts handing out the homework assignments. Armando stands by the door. He looks up at the sky. He stares at the smoke coming from the red and white striped smokestacks that rise from the power plant.

MS. SANDOVAL
Armando, that includes you too.

Armando stands still. He continues to stare. He looks at the TICKING clock. Back to the smoke. Ms. Sandoval notices, hands the assignments to a student to distribute, and walks over.

MS. SANDOVAL (CONT'D)
(gently)
Please take a seat.

ARMANDO
But I can't.

Ms. Sandoval follows Armando's eyes. She too, stares at the smoke.

ARMANDO (CONT'D)
My sister showed me a website the
other day. It looked like a map or
something.
(beat)
It used different colors to show the
air pollution. You know what color
Sun Valley was Ms.?

MS. SANDOVAL
What color, *Mando*?

ARMANDO
It was red!

MS. SANDOVAL
Red??

ARMANDO
YES! But not like any kind of red,
it was a ***dark*** red. You know what the
dark red means?

Ms. Sandoval tries to gather her thoughts. Thinking of a possible response to her student.

ARMANDO (CONT'D)
It's really bad Ms.! It means that
We have the most worst air there is.

Ms. Sandoval . . . speechless.

ARMANDO (CONT'D)
And you know the alarms we hear
every couple hours?
(beat)
Well I read it means that they are

letting bad chemicals into our air.

MS. SANDOVAL

Armando, you shouldn't worry about that stuff.

ARMANDO

What do you mean Ms.? They are doing this on purpose! They know they are making our air worse! This is why we are a dark red on the pollution map!

MS. SANDOVAL

Yes, but there isn't much we can do.

ARMANDO

Yes there is! You're our science teacher. Teach us about how our air is making us sick. How it's making our parents sick. Teach us.

Fear fills his big eyes as he registers the Power Plant's SIRENS going off. He runs inside the classroom, grabbing Ms. Sandoval's hand and pulling her inside with him. The door SHUTS behind them.

ARMANDO (CONT'D)

EVERYBODY!! HOLD IN YOUR BREATH.

Confused looks fill the room.

ARMANDO

(frustrated)

JUST DO IT!!!!!

The **SIRENS** come to an end.

Armando sighs. Silence fills the room. Ms. Sandoval approaches. Places her hand on his shoulder.

MS. SANDOVAL

(quietly)

We can keep talking after school. But now, please take a seat.

EXT. SCHOOL HALLWAY – DAY

The school bell RINGS. The hallways fill with chatty students, all rushing to get home.

ARMANDO emerges, running through the crowd. Trying to get to MS. SANDOVAL's classroom.

INT. CLASSROOM – DAY
Armando runs into the classroom. Sweat running down his cheeks.
ARMANDO
I'm here Ms.!
Ms. Sandoval, turning off her projector and closing her laptop. Turns and smiles.
MS. SANDOVAL
Yes, I can see you.
ARMANDO
So the pollution coming from that place.
MS. SANDOVAL
(giggles)
You mean the Power Plant?
ARMANDO
Yes! How do we stop it?
MS. SANDOVAL
I don't think we can.
ARMANDO
You see Ms.! No one listens. No one does anything! If my sister didn't show me, I wouldn't know either. I bet other people don't know either!
(beat)
Did you know?
MS. SANDOVAL
What do you want to do?
ARMANDO
I don't know. Tell everyone? Protest?
MS. SANDOVAL
How about . . . an assembly?
ARMANDO
Yes! I want to tell everyone in the school. But I also want to tell everyone that lives in Sun Valley. We just need to translate a bunch of things. I can use Google Translate and my mom can help me fix it. I can make flyers too! And big posters! I

                    just need the art supplies. You have
                    a bunch Ms! Can I borrow some? Also,
                    do you think it's better if we go
                    knock on peoples houses or –
                              MS. SANDOVAL
                    Okay. One step at a time. We will
                    start with our classroom and then
                    move to our school and then our
                              community.
Armando's eyes fill with sadness. He was expecting immediate change.
                              ARMANDO
                    So what do I do now?
                              MS. SANDOVAL
                    For now, you can work on posters at
                    home to put up around our school.
                    Take as many supplies as you need!
                              ARMANDO
                    As many as I need?
                              MS. SANDOVAL
                    And as many as you can carry. That
                    backpack of yours seems heavy!
Armando blushes and GIGGLES. He walks over to the supply closet.

INT. ARMANDO'S APARTMENT – DAY

Armando walks into his living room and sets everything down. His little hands struggled to hold the rolls of colorful Butcher Paper and bag of paints. His backpack barely zipped up.

He walks over to give his mother a kiss on the cheek. She greets him with a glass of water.
                              ARMANDO
                              (IN SPANISH)
                    Ma! Guess what I did? I convinced my
                    teacher to help me tell people about
                    the bad chemicals in the air coming
                    from the big red and white pipes. We
                    are going to protest! And you won't
                          be sick anymore ma!
Irmas's eyes fill with tears. She looks at her son, fondly.
                              ARMANDO (CONT'D)
                    But we're going to need your help!

                    IRMA
        Of course, mijo. If there's anything
          I can do to help, I'll do it.
  They embrace. Armando's smile and her silent tears are enough to show that his efforts have already made an impact.

23. *I Am a Drop of Water*

YAYING WU

This digital ecowriting was created by an undergraduate student in an environmental justice class at the University of California, Los Angeles (UCLA).

View "I Am a Drop of Water" via this URL or the QR code: https://tinyurl.com/2tj87r7z

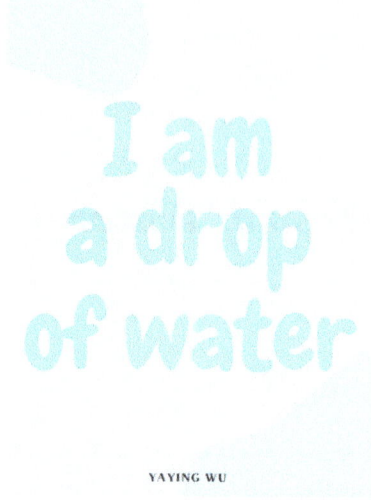

Figure 23.1: Still image from digital ecowriting "I Am a Drop of Water."

Figure 23.2: QR code for digital access to digital ecowriting "I Am a Drop of Water."

24. Dear Diary

ARBREAN SEARS

Dear Diary,
Today I had a good day. I got an entire seat to myself on the bus ride this morning! It was really foggy, or so I thought, until my bio teacher told me it was smog from the sky. I decided to stay in the classroom for lunch because the air smelled bad outside. During PE we had freeplay so obviously I chose to walk the track because I was NOT going to play basketball with the boys. They take it too seriously. After school, I went over to grandma's to check on her. She's sick. It's something about her heart I forget what it's called, but it makes a lot of things difficult for her. She had been sitting inside all day so I offered to take her outside but she said the air was too bad. Sometimes I wish she wasn't sick. A lot has changed since we found out. I stayed with her a bit longer and helped her take her medicine. After that I went outside to wait for the city bus. It felt kind of hard to breathe but I just figured I was tired. When I got home I took a nap and when I woke up, dinner was ready. Chicken noodle soup, mmm my favorite. Going to go eat now.
Talk to you soon.

Dear Diary,
Last night I had trouble sleeping. For some reason I kept waking up out of my sleep with a really bad cough, but when I woke up it was gone. When I went into the kitchen to eat breakfast my mom was watching the news in the living room. Today the air was unhealthy but only for sensitive groups. Boy do I wish I was them, then I could get out of PE. My little brother has asthma. He has

to take his medicine twice every day, and on really bad days he has to sleep with this breathing machine. My uncle says it's the air making everyone sick but my mom always says it runs in the family. That's why grandma got sick, and grandpa had his heart attack. Sometimes I wonder why our family had to have these illnesses. The sky was pretty today. It wasn't all grey like yesterday. But when I got to school there was this very stinky smell. My school is right next to a meat packing factory so the stinky smells came from the dead meat they were cleaning. Today in bio my teacher said that the meat industry causes pollution in the air. We also learned that pollution can affect our health. So maybe my uncle was right ... We ran the mile for PE today and I felt like I was going to DIE. You may think I'm overexaggerating but it was really hard to breathe after I finished and the stinky hair made it no better. Going to go take a nap now.
Smell you later!

Dear Diary,
Today I stayed home from school. I still wasn't feeling well after that mile and I had a bad cough for the rest of the evening all the way until the next morning. When I woke up, it was still hard to breathe. I told my mom and she said she was going to take me to the doctor. So the doctor tells me that I have asthma and a small form of bronchitis, which was why I've been coughing and having trouble breathing. When they told me I was sad because I didn't want to be like my brother who has to take his pump every day. My doctor prescribed me medicine and suggested that I don't do a lot of moving and that I rest ... like grandma. I felt discouraged after finding out I also had an illness because having health issues makes a lot of things hard for my little brother and grandma ... and now me too. When my mom told my uncle about it, he claimed that it was the pollution that our city has that caused me to get sick. He also said that going to school and being exposed for prolonged periods to the harmful air and smells made it even worse for me. I hope that somehow I can still get better, even though there is no cure for asthma. Got to catch up on homework!
Bye-Bye

Dear Diary,

Today was my first day back at school since I'd been sick. For now, I also have to take my inhaler two times a day and carry around another inhaler in case I have an asthma attack. I don't like that I have to do this but I have no other choice. Coincidentally, since we've been talking about the climate and pollution and stuff in my bio class, we had guest speakers from the local college come talk to us. If you haven't noticed by now, science is one of my favorite subjects. These speakers were saying the same things my uncle was saying, and they backed it up with science! So he was right. It is our air, but how can you escape from bad air so that you can heal? Well we can't because air is all around us. We either have to move out of this city or somehow fix the pollution in the air; however, that would be a hard task. According to our guest speakers, the pollution from these factories is caused by big companies, and these big companies don't really care. So in order for us to fix the air, they would have to stop their harmful methods ... which they won't do. Our speakers were part of a group that went to schools and other places and spoke about the harmful pollution in the air. They hold events and protests and even try to talk to people in higher positions in the government in order to advocate for better living conditions for us because the pollution in the air is dangerous to our health. They left their information so we can keep in contact with them and join some of their events. I think I will do that. Got to go!

Dear Diary,

When I went home, I started to do research about pollution. It affects so many things. It affects our air, which affects our climate, which affects our weather, which affects plants, which affects animals, which affects food supply, and overall affects us in so many ways! I was so surprised to be learning this all right now. I think this is something that we should have been taught. During my research I found that pollution can cause many negative health effects to our bodies. It can cause issues with the heart and the lungs, and can even lead to cancer! It's so scary and many people in my community and FAMILY suffer from these types of illnesses. So maybe it's the pollution that is harming us. I got in touch with the speakers from my bio class and they invited me to

an event they're talking at where they're informing a group of business owners how they are harming the people and the planet and how they can cut the amount of pollutants they release into the air. I hope that when I get older I can do the same because our lives are in danger and we must do something to fix it. Anyway, talk to you later!
P.E.A.C.E.

25. *To Our Beautiful Mother*

Mandie Torres

"New Beginning" by: Raskahuele (2012) (https://www.youtube.com/watch?v=adk84FsJa1c)

Spanish/Original Version
Si yo pudiera detener el tiempo
Lo ocuparía para escapar y olvidar todo lo que nos está ocurriendo
Lo ocuparía para concientizar a todos a cuidar nuestra Madre Tierra
A cuidar todo lo que nos enseña
Y entender que todos somos uno y uno somos todos . . .

Gente que va caminando firme
Víctimas de una obvia discriminacion
Viviendo lejos de casa . . .

Nuestros ancestros eternos
En el alma viven y se defendieron sin palabra
Nuestra crítica arma siempre sigue, es revolución armada. . .

English Translation
[If I could stop time
I would use it to escape
and forget everything that is happening to us
I would use it to raise everyone's consciousness to care for Mother Earth
To take care of everything that she teaches us
And understand that we are all one . . .

The people who are walking firm
Victims of obvious discrimination
Living away from home . . .

Our eternal ancestors
They live in our souls
And defended themselves without hesitation
Our critical weapon always continues, it is armed revolution]

"New Beginning" is one of my favorite songs of all time and it's by one of my favorite bands as well. I specifically love it not just because of its rhythms and sounds but also for the meaningful and powerful lyrics. In the first section they are singing about what they would do if they could stop time. There is so much oppression and injustice in the world that they wish they could escape. Then they start to sing about how they would also use that time to raise everyone's consciousness to take care of *nuestra Madre Tierra* [our Mother Earth]. When they sing about Mother Earth and all that she teaches us, I find it super powerful, and to this day, after countless times hearing this song, it still brings a passion but simultaneous grief upon me because she's been abused for far too long. I start to think about how we have to protect her and pay close attention to the ways that she teaches us.

The second section I grabbed from the song reminds me of the people who suffer from injustices, including the effects of climate change and climate migration, and how they've remained strong no matter where they're at even when having to live away from home. When we think about who these people are, it's usually Indigenous, Black, and poor communities of color who aren't even responsible for climate effects. The third and last section I included from the song speaks to our internal will and capabilities to continue fighting for climate, environmental, and social justice. Our ancestors power us and as long as we keep a critical lens, that is a form of revolution. But I want to speak more directly to you, *mi Madre Tierra* (my Mother Earth).

To our beautiful Mother,

My heart aches at the thought of how much you are hurting right now. It aches when I see the way your waters have been poisoned. The way oil and plastic have filled your veins. The fact that your fresh air has been polluted with toxins. That humans are disturbing and intruding on your body with pipelines.

I'm sorry that many greedy humans have seen you and treated you as less than. I'm sorry that they don't see the life you give us, don't see everything you teach us. I'm sorry that you have been pillaged, raped, extracted from, seen as disposable and subordinate. I'm sorry that so many have taken, taken, and taken and not given back.

I'm sorry that many of us have taken for granted everything you have to teach us. I'm sorry that at one point I was one of those people.

Pero Madre Tierra ya desperté. Te siento. Cuando veo los árboles, las nubes, los mares, las plantas, los animales, te siento en mi alma y en mi corazón. Siento tu dolor pero también siento tu poder y fortaleza. Gracias a la gente Indígena que sigue cuidándote y luchando por ti. Gracias por todo lo que nos enseñas. /

But Mother Earth I have awoken. I feel you. When I see the trees, the clouds, the oceans, the plants, the animals, I feel you in my soul and in my heart. I feel your pain but I also feel your power and strength. Thank you to the Indigenous peoples who continue to care for you and fight for you. Thank you for everything that you have taught us.

You teach us the natural cycles of life. You teach us how to be respectful and to care for one another. You teach us equality and reciprocity and how we all matter. You teach us that hierarchies are a selfish human construct and that things are not top-down nor linear. You give us literal life. You provide us with the fresh air that we need to breathe. With the water that we need to drink. With the warmth we need to survive. With the food that we need to eat.

Agua es vida, tierra es vida, lumbre es vida, aire es vida, nuestra Madre Tierra es vida. / Water is life, earth is life, fire is life, air is life, our Mother Earth is life.

And I'm so sorry that many of us have been consumed with settler-colonial and capitalist mentalities that don't allow us to see that or take it for granted. *Madre Tierra eres tan poderosa. Nos das nuestra medicina de tu tierra.* / Mother Earth you are so powerful. You give us your medicine.

One thing I have learned though is that you will live on. We will keep fighting for you but no matter what happens, you will live on. You have shown us that no one is above you, you will take your power back and continue to be life. *Seguiremos la lucha para cuidarte y protegerte. Gracias por todo lo que nos prestas y todo los que nos has enseñado todos estos años.* / We will continue the fight to care for you and protect you. Thank you for everything that you share with us and everything that you have taught us over the years.

Con mucho amor,
Una de tus varias hijas
With much love,
One of your various daughters

Reference

Raskahuele. (2012). New Beginning [Song]. In *New Beginning*. Raskahuele.

26. Dear Councilwoman Monica Rodriguez

NICOLE HALL

Dear Councilwoman Monica Rodriguez,

My name is Nicole Hall and I am a resident of Tujunga. I am writing in hopes that district 7 can adopt more eco-friendly practices and add more opportunities for residents to connect with the natural world.

As a city with a high percentage of apartments, many residents – myself and my family included – often lack the resources needed to experience nature in its original form. On my block, there are seven apartment buildings and two residential care homes. None of these living areas have sufficient grass, trees, or other plants. Aside from the small local park, there are few locations available for children to play safely. I am writing to request the addition of a new park in the neighborhood. Research has shown that parks and other open spaces have numerous positive effects on communities including improving water and air quality, bettering mental and physical health, and increasing local tax and property value. Every day as I walk my brother to school, we pass a large empty lot that is surrounded by a worn-down chain-link fence and is typically full of trash. This space could be repurposed into a beautiful park for children to play and neighbors to gather.

In addition to creating more park space, I ask that we implement side-walk gardens that other local cities have created. Pacoima Beautiful, an environmental justice organization in Pacoima, had success with building a garden along an area that was once an illegal dumping site. As an apartment-heavy neighborhood, there is constantly old furniture and other trash lining our streets. Gardening has many positive benefits and because of the lack of space, most residents are not able to garden. A side-walk garden can help limit the illegal dumping while also allowing children, families, and other residents to get experience gardening and growing their own food and plants.

I ask that Tujunga offer a weekly or monthly farmers market. In the city of Tujunga, there are no health food markets and given that many residents do not have the ability to garden, much of the neighborhood is not able to access fresh fruits and vegetables. Wealthier nearby cities such as La Crescenta have weekly farmers markets where locals can purchase foods and goods from small businesses. Adding a farmers' market in Tujunga can help small businesses earn more money while also giving us residents an opportunity to purchase fresh and healthy foods.

Finally, I ask that the city offer more educational opportunities for residents to learn about the importance of environmental justice and protecting the environment. Because the library is located next to the park and municipal building, there are ample opportunities for the city to partner with the library and community to help educate children and adults on important topics related to the climate crisis and other ways that the natural world can help better a person's life. Throughout my K-12 education, I did not have any opportunities to learn about environmental-related issues, and by offering free classes to the community, you can help foster a more environmentally aware community so that we can help preserve the environment.

Thank you for considering these changes,
Nicole Hall

Part IV: Resources for Ecowriting

27. Introduction to Ecowriting Lesson Units

SYDNEY RICHMOND AND ANDREA GAMBINO

The following units are designed for educators who are interested in implementing ecowriting in their classrooms. Each unit guide contains an overview, guiding questions, learning outcomes, target vocabulary, a daily pacing plan with embedded texts, and multimedia presentations with supplementary documents. Lessons should be adapted to best meet the needs and interests of the students in order to be developmentally appropriate, culturally relevant, and connected with course objectives. To modify any of the Google Presentations, Jamboards, Documents, etc., click *file, make a copy, and save them to your local Google Drive*. Remember to edit the share settings so that students can access the materials.

Unit 1: Exploring Our Relationships with Nature, focuses on providing student-centered ecowriting experiences to guide students as they think about their relationships with the natural world. Structured as a five-day unit, students begin their adventure with ecowriting by participating in a nature walk. To help make sense of their experiences in nature, students are encouraged to be mindful of how their senses engage with nature by collecting audio recordings, taking photographs, or producing their own sketches. Through disrupting the binary of biophilia (love of nature) and biophobia (fear of nature), students consider the complex relationships that exist between themselves, their peers, and the environment. Additionally, students engage in genre studies, with a particular emphasis on poetry, to analyze how various ecowriters discuss their relationships with the ecosystems that surround them before creating their own blackout poetry of Amanda Gorman's "Our Purpose in Poetry, Or Earthrise." As a culminating experience, students craft their own digital stories, culling from their ecowriting analyses and creations during this unit. When students consider their relationships with the natural world through ecowriting, this can serve as an entry point for more critical analyses (e.g., featured in Unit 2: Greenwashing) about large-scale issues impacting environmental sustainability. All recommended texts, scaffolded activities, Google Jamboards and Documents, are provided in the Unit Guide and are embedded in the lesson sequence. This unit is tailored to upper elementary and young adolescent learners but could be adapted for all ages.

Unit 2: Greenwashing – Disrupting False or Misleading Claims of Environmental Ethics, analyzes how advertising techniques mislead or falsely assert ethical practices that advance environmental sustainability. This 10-day unit draws upon the Critical Media Literacy Framework: Conceptual Understandings and Guiding Questions (Kellner & Share, 2019, p. 8), as a tool to support students as they develop and expand critical stances about persuasive advertising techniques, particularly how some companies engage in greenwashing. Students participate in several critical media analyses to increase their awareness and critical thinking about media and greenwashing. Moving from critical media analysis to production, students form inquiry-based learning teams to design their own research question they are curious about related to greenwashing (e.g., fast fashion, animal testing, etc.). Guided by scaffolded project check points, students engage in the full process of inquiry, including collecting their own research, synthesizing their findings in written, oral, and multimedia formats, and producing their own ecowriting countermedia that challenges issues related to greenwashing. After sharing their projects with their communities, students can engage in an additional extension lesson to reflect on what they learned through a Socratic seminar. This can help them continue their ecowriting with a letter-writing campaign to a company or organization that is engaging in positive environmentally sustainable practices.

All scaffolded activities, Google Presentations (including embedded links for recommended texts), and supplementary materials are provided in the *Unit 2: Instructional Resources Guide*. This guide also supports educators to deepen their understanding of critical media literacy, inquiry-based learning, and analyzing source materials with students. This unit is tailored for secondary and post-secondary students but could be adapted for upper-middle grades learners and beyond.

References

Climate Reality. (2018, December 4). *24 hours of reality: "Earthrise" by Amanda Gorman* [Video]. YouTube. https://www.youtube.com/watch?v=xwOvBv8RLmo&t=1s

Kellner, D., & Share, J. (2019). *The critical media literacy guide: Engaging media and transforming education.* Brill/Sense Publishers.

28. Ecowriting Unit 1: Exploring Our Relationships with Nature

SYDNEY RICHMOND AND ANDREA GAMBINO

*Digital unit version available at: https://tinyurl.com/dybxnc8v

Figure 28.1: Scan this QR code for access to the digital version of this unit.

Unit Overview: In this five-day unit plan, students explore their relationship with nature. They begin by reflecting on their experiences and relationships with the natural world by engaging in a nature walk. Next, they examine multiple forms of media to consider how others use ecowriting as a tool to express their thoughts and beliefs about nature. They also explore and discuss how to go beyond binary thinking about an individual's possible biophilia and biophobia. Last, students construct individual digital stories to describe their own relationship with nature and to inspire others to engage with ecowriting as a tool of reflection, connection, and action to respect the natural world and all living things.

Guiding Questions:
- What is my relationship with nature?
- What is biophilia?
- What is biophobia?
- How can I use ecowriting to explore my relationship with the natural world?

Learning Outcomes:
- Discover and reflect upon relationships with nature.
- Explore and analyze biophilia and biophobia.
- Engage in ecowriting through media analysis and production.
- Share ecowriting to inspire others to deepen their relationships with nature.

Target Vocabulary:
- **Biophilia:** An individual's connection with and appreciation of nature, the natural world, and the ecosystems that surround them.
- **Biophobia:** An individual's disconnection or discomfort from elements of the natural world and/or factors in ecosystems.
- **Ecowriting:** Print and non-print (video, audio, photography, multimodal, multimedia, other forms of digital communication) writing to engage in reflection and storytelling about the environment, ecosystems, and our relationships with the natural world.

Day 1: My Relationship with Nature

Lesson Opener: *Observing Life Around Us*
- Have students think about *how* they observe the people and things around them. Ask students: *What are ways that you collect observations about the world around you?*
- Instruct students to silently watch this video and follow along with the guided instructions: Simons, D. (2010, April 28). *The monkey business illusion* [Video]. YouTube. https://www.youtube.com/watch?v=IGQmdoK_ZfY
  - After watching the video: Ask students for their reactions and if they were able to count the correct number of ball passes (e.g., 16), whether they observed the gorilla, and if they noticed the change in color of the curtain behind the teams.
- Have students reflect and share the ways they engage with the natural world and observe their surroundings. Ask: *What do you often notice when you are in nature?*

Activity 1: *Discovering How I Define My Own Relationship with Nature*
- Tell students that they will go on a nature walk around their school community. During the nature walk, they should find a space where they can sit silently for a while. As they sit silently, they should focus on their senses and think about how they are engaging with the natural world (anything that is not created by humans, e.g., dirt, plants, insects, water, clouds, etc.).
- Encourage students to document their experience by writing or drawing. Provide guiding questions to scaffold their writings and drawings, such as: *What do you see, hear, smell, taste, feel, touch?* Ask students to take notes, photos, and recordings about their observations to describe how their senses are engaging with the natural world.
  - *Note: They will need to bring a notebook, pencil/pen, mobile device or tablet (to audio record sounds and/or take pictures), and any drawing materials they would like to bring to sketch their surroundings. Students can draw/sketch (images, sounds, landscapes, colors, etc.) and write any observations that can help them think about their observations and relationship with nature.*

Lesson Closer: *Ecowriting Your World*
- Return to the classroom from the nature walk. Explain to students that they are engaging in a process called: *ecowriting*. Ask students to synthesize their observations to generate a reflective ecowriting paragraph. Students may create captions for their favorite drawings and photographs to help tell their story about their relationship with nature. Provide guiding questions, such as: *What did you notice about nature? How did you feel being in nature? During your nature walk, what made you comfortable and/or uncomfortable?*
  - *Note: Ecowriting can be print and non-print (video, audio, photography, multimedia, and any form of communication) writing to engage in reflection and storytelling about the environment, ecosystems, and our own relationships with the natural world. If students need more time to finish, ask them to complete their ecowriting for homework and/or expand this activity into an additional instructional day.*

Day 2: *Sharing About Our Relationships with Nature*

Lesson Opener: *Analyzing Lindi Nolte's "A Love Poem to Our Earth"*
- Invite students to view this video: Nolte, L. [Tedx Talks]. (2020, April 6). *A love poem to our Earth* [Video]. YouTube. https://www.youtube.com/watch?v=rWR86_YODaU

- Have students discuss with a partner how Lindi Nolte discusses her relationship with nature. Provide guiding questions, such as: *How does Lindi Nolte use descriptive and sensory language to describe her relationship with nature? What are your reactions and takeaways?* Encourage students to share with the whole class what they discussed with their partners. Help students build connections with their observations and their ecowriting from Lesson 1.

Activity 1: *Let's Jam!: Sharing Our Ecowriting About Our Relationships with Nature*
- Use or adapt this Google Jamboard: https://tinyurl.com/44mmddha
  - Have students prepare to share their ecowriting. Invite them to add their writing, drawings, and photographs to their own Google Jamboard slide.
  - Next, number students off to form small groups to share their ecowriting with each other. Ask students to discuss the similarities and differences in their nature walk and ecowriting process. Remind them to honor and respect each other's different perspectives. With the whole class, encourage students to share their different experiences. If a student mentions a peer's ecowriting, request permission from the other student, and if granted, display it to the class to provide visuals.
  - *Note: To modify the Google Jamboard: log into your Google account, click the above Google Jamboard link, click the three dots in the top right-hand corner, and then click make a copy and save it to your local Google Drive (change the share settings so that anyone can edit). This activity could also be adapted as a gallery walk or show and tell format where students read their ecowriting and show their original drawings/sketches and photographs. If this lesson is used in-person, students could spread out with their groups and play their sound recordings while they share their writing. Or, if this lesson is used in an online format, students could join breakout rooms based on their group numbers.*

Lesson Closer 3-2-1 Summarizer
- Ask students to complete a 3-2-1 summarizer to reflect upon their own ecowriting experiences and what they learned from their peers. For example, ask students to write on paper:
  - *3 key takeaways* from their nature walk, ecowriting process, and/or discussions.
  - *2 questions* they have about their own/peer's relationship with nature.

- ○ *1 personal connection* about how their thinking about their relationship with nature was affected from the nature walk and ecowriting.

Day 3: Beyond the Binary: Biophilia & Biophobia

Lesson Opener: *Comfort and Discomfort with Nature*
- As a class, construct a t-chart (or print out this graphic organizer from NCTE's *Read, Write, Think:* https://www.readwritethink.org/classroom-resources/printouts/chart) to describe how individuals may be comfortable and/or uncomfortable when experiencing different elements in the natural world. Model an example, such as: *Individuals might feel comfortable going in a group for a nature walk around their school, but could be uncomfortable going for a hike alone in the mountains.*
- After students are finished sharing their ideas, discuss the importance of recognizing that we all have different relationships with the natural world and to not judge others' feelings, fears, or thoughts they may associate with nature.

Activity 1: *Beyond the Binary: Biophilia and Biophobia*
- Ask students how they would describe the word *binary*. As students provide their answers, discuss that a binary is an either/or construction, often suggesting that there is one correct or absolute interpretation, feeling, or way of being. Engage in dialogue with students about the importance of going beyond binary thinking, as it can suggest a deficit-mindset or an either/or mentality when considering our own or others' relationships with nature. Next, discuss the following target vocabulary definitions for biophilia and biophobia.
  - **Biophilia:** An individual's connection with and appreciation of nature, the natural world, and the ecosystems that surround them (love of nature).
  - **Biophobia:** An individual's disconnection or discomfort from elements of the natural world and/or factors in ecosystems (fear of nature).
- Have students share their ideas about how we can go beyond binary thinking when considering our own and others' biophilia and biophobia. For example, the teacher could pose: *I like going to the beach but I do not feel comfortable swimming in the ocean.* This example reveals my biophilia of the beach as well as my biophobia of the ocean.

Lesson Closer: *Collaborative Definitions*
- Invite students to build a collaborative definition that describes how individuals might experience biophilia and biophobia simultaneously. Encourage students to share their ideas and compile them on a collaborative whiteboard (https://webwhiteboard.com/) or physical whiteboard using different color dry erase markers or sticky notes.

Day 4: Ecowriting Genre Studies

Lesson Opener: *Analyzing Photography: "Earthrise"*
- Use or adapt this Google Jamboard: https://tinyurl.com/2d8b6cvb
  - Use the Analyzing Photography: "Earthrise" Google Jamboard to show astronaut William Anders' (1968) photograph, taken from Apollo 8 as it rounded the dark side of the moon for the fourth time. Explain that when *Life* magazine published the photo shortly after it was taken, U.S. poet laureate, James Dickey wrote a poem that was printed with the photograph. Read the short poem aloud:
    And behold
    The blue planet steeped in its dream
    Of reality, its calculated vision shaking with the only love.
  - Have students pause and examine the photograph and Dickey's poem. Next, ask students to write their own 3-line poem to describe and build connections to the photograph. Invite them to add their poem to their own Google Jamboard slide.
- *Extension:* Students can read, explore, and discuss:
  - Global Oneness Project. (2022). *Earthrise: The image that shared our world.* https://www.globalonenessproject.org/library/films/earthrise
  - Moran, J. (2018, December 22). *Earthrise: The story behind our planet's most famous photo.* The Guardian. https://www.theguardian.com/artanddesign/2018/dec/22/behold-blue-plant-photograph-earthrise
  - Time. (2016, November 30). *Earthrise: The story behind William Anders' Apollo 8 Photograph* [Video]. YouTube. https://www.youtube.com/watch?v=Pu7NUQEHfe4

Activity 1: *Amanda Gorman's "Our Purpose in Poetry, Or Earthrise"*
- View Amanda Gorman's "Our Purpose in Poetry, Or Earthrise." Ask students to observe how the images, sounds, and lines of Gorman's poem communicate a story about the importance of one's

relationship with nature, considering our responsibilities for taking care of the ecosystems that surround us, and how we might take action in our local communities to contribute to environmental sustainability.
    - Climate Reality. (2018, December 4). *24 hours of reality: "Earthrise" by Amanda Gorman* [Video]. YouTube. https://www.youtube.com/watch?v=xwOvBv8RLmo&t=1s
- After viewing the video, have each student make a list of five words (e.g., nouns, verbs, and/or adjectives) that they heard or felt resonating from Gorman's poem. Next, invite them to contribute their words by reading them aloud and writing them on a whiteboard or by creating a class word cloud.
    - Mentimeter (free word cloud generator): https://www.mentimeter.com/features/word-cloud
    - Word Art (free word cloud creator): https://wordart.com/
- Guide students to observe the patterns, synonyms, and different perspectives that they had to Gorman's poem. Continue engaging the class in a dialogue and encourage students to share about the words they chose to describe Gorman's poem, their key takeaways, and reactions.

Activity 2: *"Earthrise" Remix Blackout Poetry*
- Explain to students how they will create their own blackout poems using Amanda Gorman's "Our Purpose in Poetry, Or Earthrise." Students should receive copies of the Unit 1: Day 4 – Activity 2: "Earthrise" Remix – Blackout Poem document: https://tinyurl.com/3y5bxrj8
    - *Note: To have students create blackout poems digitally, make a copy of the document above by clicking file, make a copy, and save it to your local Google Drive. Students could also use this same process to create their own local copies.*
- View the following video with students to help them understand the process of blackout poetry:
    - Kleon, A. (2015, September 21). *How to make a newspaper blackout poem* [Video]. YouTube. https://www.youtube.com/watch?v=wKpVgoGr6kE
- After viewing, check for students' understanding of the process of creating a blackout poem. Let students know that blackout poems can be created with all types of genres. Show students additional examples of blackout poems that combine drawings, colors, and other forms of creative expression, for example:

- ○ Creative Yatra. (2019). *Time for society to struggle through the dark* [Blogpost]. https://creativeyatra.com/wp-content/uploads/2019/05/Blackout-Poetry-Workshop-2.jpg
- Next, provide time for students to create their own blackout poems digitally or using the printed handouts. Make sure students have access to black markers to mark out words they would like to omit. Provide other markers, colored pencils, etc. and encourage them to also create drawings to express their relationship with nature. Provide guiding questions, such as: *How would you describe your relationship with the natural world? How do you think individuals and collectives can contribute to an environmentally sustainable planet? What message would you like to convey to readers through your blackout poem?*
- *Extension:* Expand this activity to help students think more deeply about the power of amplifying one's own voice through poetry and other forms of storytelling. Listen and discuss:
  - ○ TED. (2021, January 20). *Amanda Gorman: Using your voice as a political choice* [TED video]. YouTube. https://www.youtube.com/watch?v=zaZBgqfEa1E

Lesson Closer: *Sharing Our Blackout Poems*
- If possible, position your classroom seating arrangement to form a circle. If you are teaching in an online context, make sure the view is in *gallery-mode* so that everyone can see one another. Invite students to take turns sharing their blackout poems. Engage in full poetry jam format, inviting students to celebrate their peers' ecowriting. After each person finishes sharing, debrief their process of remixing Gorman's poem through blackout poetry. Ask them to share their key takeaways, asking: *How can we use our voices through ecowriting to reflect upon our relationships with nature and actions we can take to become stewards of the environment around us?*

Day 5: Ecowriting Our Worlds: Digital Story Production

Lesson Opener: *Class Dialogue – The Power of Storytelling*
- Guide students to think about characteristics of powerful stories they have heard and that have resonated with them. Provide guiding questions, such as: *Can you remember a story that you have heard or watched that was meaningful to you? What characteristics can you remember that made the story resonate with you?*
- Encourage students to consider features such as narration, dialogue, sound, music, voice inflection, tone, etc. If students mention a story

that they watched, ask about what images, colors, and other visuals they remember that helped illustrate the story.

Activity 1: *Construct Your Own Digital Story*
- Invite students to reflect upon what they have learned about their relationship with nature. Have them freewrite on paper for 5 minutes. Ask them to describe how their thinking about their relationship with nature has evolved during the different readings and activities they have completed during this unit. Suggest that students use their reflections to spark ideas to create a digital story.
- Guide students to brainstorm ideas to create a short digital story about their relationships with nature. Have them draft a script of the narration or dialogue they would like to use as a voiceover for their digital story. Provide six index cards, markers, and colored pencils that they can use to create illustrations to help tell their story about their relationship with nature. Students could also cut out some of their original drawings/sketches from their nature walk during lesson 1. It is best for students to draw on the index cards horizontally, this way their artwork will fit the horizontal layout for most TV or computer screens.
  - *Note: Students can compile their digital stories using six Google Slides, PowerPoint Slides, or any multimedia application to tell their story. Make sure students include a title for their story and their name on the opening slide (Google Slides: https://www.google.com/slides/about/). Have students take photos of their drawings using a mobile device or a tablet. They should save the image files on the device they will use to complete their digital stories (remind students to label their files in the order they would like them to appear on the Google Slides). Next, students can record their audio narration using a mobile device, tablet, or desktop/laptop computer via an audio recorder or voice memos app or they can use this free audio-recording tool (https://vocaroo.com/). Have students upload their audio files on the same device they will use to complete their digital stories. Students can upload all of their files from their desktop to their own Google Slides by clicking (insert image OR insert audio).*
  - *Note: Depending on students' progress, make adjustments to the number of instructional sessions you would like to provide. If students are contributing to this production outside of class, the time needed for digital media production will vary. Adjustment of project time and requirements may vary for age-levels, technology access, and learning styles.*

Activity 2: *Share-Out*
- Invite students to share their digital stories with the class. Celebrate and discuss each other's relationships with nature. Students can provide feedback to their peers using the format of *Two Stars and One Wish* (see this free printable template from *Pearson*: https://www.english.com/blog/self-assessment-young-learners/two-stars-and-a-wish/). This strategy asks students to write two positive comments about things they liked and one item of constructive advice for something that could be improved.

Lesson Closer: *Exit Ticket*
- Ask students to complete a brief exit ticket on paper. Provide the guiding question, *How can I continue to use ecowriting to explore my relationship with the natural world?*

29. Ecowriting Unit 2: Greenwashing – Disrupting False or Misleading Claims of Environmental Ethics

SYDNEY RICHMOND AND ANDREA GAMBINO

*Digital unit version available at: https://tinyurl.com/3ps6xh44

Figure 29.1: Scan this QR code for access to the digital version of this unit.

Unit Overview: In this 10-day unit plan, students explore greenwashing. First, they examine positive and negative examples of media advertisements that generate a public-facing image of environmental sustainability. Students investigate greenwashing by analyzing how false or misleading messages attempt to convince the public that a company is implementing sustainable environmental practices. Next, students engage in collaborative inquiry-based ecowriting to create countermedia that generates awareness about greenwashing. Last, students share their ecowriting countermedia with their families, friends, and communities. Students can also continue their ecowriting by creating letters of support to companies and organizations that engage in ethical environmental practices (see extension lesson).

Greenwashing – Instructional Resource Guide: https://tinyurl.com/3hwfp3u9

Guiding Questions:
- How do media and advertising impact our everyday lives?
- What is critical media literacy?
- What are the differences between dominant media and countermedia?
- What is greenwashing?
- What advertising techniques promote greenwashing?
- How does greenwashing negatively impact progress toward environmental sustainability?
- How can ecowriting countermedia productions hold companies accountable?
- How can ecowriting showcase support for ethical environmental practices?

Learning Outcomes:
- Consider how media and advertising impact our everyday lives.
- Understand and apply the tools of critical media literacy.
- Analyze the differences between dominant media and countermedia.
- Explore and analyze greenwashing.
- Examine greenwashing techniques using common texts and commercial media.
- Engage in inquiry-based research to investigate student-selected greenwashing topics.
- Create and share ecowriting countermedia that demonstrates positive and negative media representations of environmental sustainability.
- Engage in ecowriting to showcase support of ethical environmental practices.

Target Vocabulary:

- **Critical Media Literacy**[1]: An inquiry-based process for analyzing and creating media by interrogating the relationships between power and knowledge. It questions representations of class, gender, race, sexuality, and other forms of identity and challenges unjust media messages. Critical media literacy celebrates positive representations

---

[1] This is based on the definition of critical media literacy generated by the 2021 Critical Media Literacy Conference of the Americas Steering Committee, viewable here: https://tinyurl.com/t29f89u8.

and beneficial aspects of media while challenging problems and negative consequences, recognizing media are never neutral.
- **Dominant media:** Popular or mainstream media that are often commercial and tend to privilege upper- and middle-class White, heterosexual men over historically marginalized groups, people, and ideas.
- **Countermedia:** Alternative media that provide a different perspective than dominant media; often created in response to mainstream media in an attempt to challenge under-representations and misrepresentations.
- **Greenwashing:** Public-facing information that makes false or misleading claims of environmental sustainability, often promoting commercial companies or organizations through advertisements.
- **Environmentally friendly:** Environmentally friendly or eco-friendly are terms often used to describe whether a product is *earth-friendly* or *not harmful to the environment*. It commonly applies to products that contribute to green-living and practices that aid in resource conservation (e.g., energy or water) and environmental sustainability.
- **Environmental Sustainability:** Responsibility to conserve natural resources, act in harmony with the natural world, positively impact local and global ecosystems, and contribute to physical and ecological health and well-being.
- **Eco-labels:** Product branding or labeling that suggests environmentally friendly practices.
- **Ecowriting:** Print and non-print (video, audio, photography, multimodal, multimedia, other forms of digital communication) writing to engage in storytelling about the environment, interrelated ecosystems, and relationships with the natural world.

Day 1: Introduction to Critical Media Literacy and Commercial Media/Advertising

Lesson Opener: *Logo/Brand Alphabet*
- Use or adapt this Google Presentation: https://tinyurl.com/62rvjr3k
  - Slide 1: Have students guess what each brand or logo represents based on the first letter. Guide students to reflect on why they were able to name so many products and brands. Ask students, *Can you find a brand or logo that is selling something, such as: products, ideas, or something else?* Encourage students to consider who/what influences their purchases and everyday lives.

- Note: *When in presenter mode, click once after students guess each letter to reveal answers on slide 1 (see full answer key for the teacher on slide 2).*

Activity 1: McDonalds' Advertisement Analysis
- Use or adapt this Google Presentation: https://tinyurl.com/4vb97pap
  - Slide 1: Invite students to freewrite/discuss the McDonalds' advertisement. Ask: *Why do you think that someone would purchase food from McDonalds, based on the advertisement?*
  - Slide 2: Have students share their ideas and generate a class list.

Activity 2: What is Critical Media Literacy (CML)? How can we use the CML questions?
- Use or adapt this Google Presentation: https://tinyurl.com/5wzzkx4a
  - Slide 1: Have students describe the term *literacy*. Discuss that traditional *literacy* includes *reading, writing, speaking, and listening* using print-based texts.
  - Slide 2: Explain that critical media literacy (CML) *expands our understanding of literacy (reading, writing, speaking, and listening) to include images, sounds, advertising, social media, popular culture, as well as print-based texts.*
  - Slide 3: Describe that CML *deepens our abilities to critically analyze and create media by interrogating the relationships between power and knowledge.*
  - Slide 4: Express that CML *inspires critical thinking and inquiry-based questions about representations of class, gender, race, sexuality, and other forms of identities, and challenges unjust media messages.*
  - Slide 5: Share that CML *celebrates positive representations and beneficial aspects of media while challenging problems and negative consequences, recognizing media are never neutral.*
  - Slide 6: Introduce the Critical Media Literacy Framework: Conceptual Understandings and Guiding Questions[2] (PDF-versions available below):

---

[2] The Critical Media Literacy Framework: Conceptual Understandings and Guiding Questions is available on p. 8: Kellner, D., & Share, J. (2019). *The critical media literacy guide: Engaging media and transforming education.* Brill/Sense Publishers. 10.13140/RG.2.2.32448.79360 Framework available in English, Spanish, Mandarin, and Portuguese as free-to-use shareable PDFs (see instructional resource guide for Unit 2: Day 1, Activity 2).

- English: https://tinyurl.com/4v5ndatx
- Spanish: https://tinyurl.com/2fdaz8up
- Mandarin: https://tinyurl.com/mpfveh85
- Portuguese: https://tinyurl.com/mwrkf9sz
  - Slide 7: Have students revisit the McDonalds' advertisement from Activity 1. Guide them as they analyze the advertisement using the six CML questions.

Activity 3: *Dominant and Countermedia*
- Use or adapt this Google Presentation: https://tinyurl.com/a8v752ea
  - Slide 1: Have students describe the term *dominant media*. Describe that dominant media are typically *popular or mainstream media that are often commercial and tend to privilege upper- and middle-class White, heterosexual men over historically marginalized groups, people, and ideas.* Example explanation found in slide 1 (presenter notes).
  - Slide 2: View: Dissolve. (2014, March 21). *This is a generic brand video* [Video]. YouTube. https://www.youtube.com/watch?v=2YBtspm8j8M. Ask students: *What could Dissolve be suggesting about some of the common techniques of advertising? How could these techniques in dominant media advertising affect or position consumers?* Invite them to discuss their observations/ideas with a partner or debrief as a full class.
  - Slide 3: Invite students to share what they think of when they hear the term *countermedia*. Explain that *countermedia are alternative media that provide a different perspective than dominant media, often created in response to dominant media in an attempt to challenge under-representations and misrepresentations.* Show the examples of countermedia (slide 3).
  - Slide 4: Have students talk with a partner or in a small group about the countermedia examples on slide. Provide guiding questions to help facilitate their discussion, such as: *How do the countermedia examples impact your analysis of the previous dominant media McDonalds' advertisement? What do you notice about the kinds of words, images, or colors in the countermedia examples? How do they differ from the dominant media example? What kinds of ideas are being shared differently in the countermedia examples as opposed to the dominant media example?*
  - Slide 5 *(Extension):* Invite students to create their own countermedia. Re-create a media advertisement (on paper or digitally)

to challenge what is represented, misrepresented, or underrepresented in the dominant media advertisement that they recalled or observed.

Lesson Closer: *Check for Understanding*
- Check for students' understanding of the differences between dominant media and countermedia. Have partner groups share their observations and ideas (encourage them to go deeply by discussing examples from class or from their daily lives).

Day 2: Introduction to Greenwashing

Lesson Opener: *Greenwashing K-W-L Chart ("I know . .")*
- Use or adapt this Google Presentation: https://tinyurl.com/nfvakdrf
  - Slide 1: Invite students to express what they **know** about greenwashing by building connections to their prior knowledge. Have students draw a three-column K-W-L chart on paper or print slide 1 as a PDF. Students should complete only the "I know" section at this time. Provide the prompt: *Describe a product that you think is NOT environmentally sustainable. Explain how the producers of this product have created a false or misleading impression that this product is environmentally safe.*
  - Have students share their writing with the whole class or with a partner.

Activity 1: *"A Fiji Water Story" and Greenwashing Techniques*
- Use or adapt this Google Presentation: https://tinyurl.com/un8956uu
  - Slide 1: Show the video on slide 1: Our Changing Climate. (2017, July 28). *Greenwashing: A Fiji Water Story* [Video]. YouTube. https://www.youtube.com/watch?v=mOpa8kd6fBI. After viewing the text, have students freewrite for 5 minutes to respond to the following prompts: *What is greenwashing? What kinds of messages were shared by Fiji's advertising that represent elements of greenwashing?*
  - Slide 2: Discuss students' thoughts and generate a class list about elements of greenwashing.

Activity 2: *Greenwashing K-W-L Chart ("I want to know . .")*
- Use or adapt this Google Presentation: https://tinyurl.com/ddbhus25

- Slide 1: Have students revisit their K-**W**-L charts. Invite students to answer: *What do you want to learn and explore about greenwashing and its impact on the environment, consumers, and our daily lives?*
- Slide 2: Encourage students to share their freewrites and create a list of students' interests on slide 2.

Activity 3: *Greenwashing Reading Groups*
- Use or adapt this Google Presentation: https://tinyurl.com/24yeyze9
  - Slide 1: Have students work in small reading groups to explore one or more texts related to greenwashing and use the CML Guiding Questions to analyze the text(s).
  - Slide 2: Refer students to the CML Guiding Questions (printable CML Guiding Questions PDFs available in the presenter notes).
  - *Note: Encourage students to discuss which text(s) they would like to examine with their peers. For younger learners (K-5), it could be helpful to read/view a single text aloud as a whole class and work together to answer just one or two of the CML guiding questions at a time.*
  - Slide 3: Text 1: RDC Global. (2020). Six Sins of Greenwashing [Infographic]. https://www.rcgdglobal.com/wp-content/uploads/2020/07/105999832_10158242112895610_4270567386174372578_o.jpg
  - Slide 4: Text 2: Mishra, S. (2016, March 23). *Case study: A bottle water brand, an ethical obligation, and everything in between.* Medium. https://medium.com/@swapnilmishra/a-water-bottle-brand-an-ethical-obligation-and-everything-in-between-3bfcf8e568c2
  - Slide 5: Text 3: Walker, R. (2008, June 1). *Consumed Water Proof.* The New York Times Magazine. https://www.nytimes.com/2008/06/01/magazine/01wwln-consumed-t.html
  - Slide 6: Text 4: Osman, J. (2020, November 2). *Greenwashing: The Tipping Point.* ClientEarth. https://www.clientearth.org/latest/latest-updates/stories/greenwashing-the-tipping-point/
  - Slide 7: Text 5: Johnson, M. J. [TEDxSkift]. (2020, November 16). *Guide against greenwashing* [Video]. YouTube. https://www.youtube.com/watch?v=5AUDasE1h1k
  - Slide 8: Text 6: Acaroglu, L. (2019, July 8). *What is greenwashing? How to spot it and stop it.* Medium. https://medium.com/disruptive-design/what-is-greenwashing-how-to-spot-it-and-stop-it-c44f3d130d5

○ Slide 9: *Extension Activity*: Ask students to view:
  ▪ Kierbel, A. (2021, July 15). *ECOMEDIA* [Video]. YouTube. https://www.youtube.com/watch?v=JUbFJwqzPlQ
  ▪ Invite students take notes about what they observe about how the creator(s) of this video use different genres, images, sounds, visuals, and colors to help convey their message about how media (e.g., cell phones) produce harmful impacts on the environment. Ask students to review and apply the six CML Framework Guiding Questions as they create observations about how this video is constructed. Discuss your observations/responses as a class.

Lesson Closer: *Greenwashing K-W-L Chart ("I learned..")*
- Use or adapt this Google Presentation: https://tinyurl.com/7y5prmkk
  ○ Slide 1: Have students revisit their K-W-L charts. Ask students: *What did you learn about greenwashing and its impact on the environment, consumers, and our daily lives? What details or clues did you discover about how to identify greenwashing in advertisements?*
  ○ Slide 2: Invite students to share and create a list of key takeaways.

Day 3: Environmentally Friendly Practices, Environmental Sustainability, and Eco-labels

Lesson Opener: *Defining Environmentally friendly, Environmental Sustainability, and Eco-labels*
- Use or adapt this Google Presentation: https://tinyurl.com/dm8reh3t
  ○ Slide 1: Invite students to respond to the guiding questions: *How would you define or describe the meaning of a product that is environmentally friendly? How would you define or describe how an individual or group can contribute to environmental sustainability? How would you define or describe the purpose of an eco-label? Can you think of any examples? If so, draw it or explain it.*
  ○ Slide 2: Have students talk with a partner or in a small group to share their ideas/responses to the guiding questions (slide 1). Next, synthesize students' definitions and examples relating to the terms, *environmentally friendly, environmental sustainability, and eco-labels* using the three-column list (slide 2).

Ecowriting Unit 2

- Slide 3: Review the additional definitions of environmentally friendly, environmental sustainability, and eco-labels. Encourage students to observe the similarities and/or differences they notice between the definitions. Next, have students examine the *Crema Joe* photograph on slide 3 and answer the following questions: *What do you notice about the colors and eco-labels on their "reusable capsule" coffee filter? Could this product be considered environmentally friendly? Does it contribute to environmental sustainability? Do you observe an eco-label? If so, how?*
- Slide 4: View the Crema Joe video: Crema Joe. (2019, March 10). *Reusable coffee pods – making a difference with every brew* [Video]. YouTube. https://www.youtube.com/watch?v=MYlsQvUH_to&t=55s. Ask students to observe how Crema Joe contributes to environmentally friendly practices, environmental sustainability, and accurate eco-labeling.

Activity 1: *Conscious Consumption Product Mapping*
- Use or adapt this Google Presentation: https://tinyurl.com/vphh3vtk
  - Slide 1: Share the example on slide 1 (in presenter mode) to provide a visualization of all of the elements in a McDonalds' Happy Meal. Ask students: *Did you observe any elements or materials that are or are not environmentally friendly?* Next, encourage students to think about one commercial product and draw it. Have students create a concept map on paper or on a digital concept-mapping tool (such as: Bubbl.us) that shows all the elements or materials that they think have gone into the development of this product. Ask probing questions: *How is this product made? Is it plastic? Is it metal? Is it BPA free?* (To explore BPA free with students, visit: https://www.greenmatters.com/p/bpa-free).
    - Note: *This concept map was created by Vivian Vasquez (2014) and her preschoolers when they were examining all the items that go into creating a McDonalds' Happy Meal. Vasquez and her preschoolers discovered that there are many people who made lots of choices to create a Happy Meal. The example above also connects with CML Guiding Question 1: Who are all the possible people who made choices that helped create this text?*
  - Once students are done drawing their concept map to showcase how a commercial product is constructed, ask them to engage in a turn and talk with a peer to discuss the products they deconstructed in their concept maps. Encourage them to ask each other

probing questions, such as: *Is there a logo that suggests that the product is environmentally friendly? How do you know? Are there any colors or symbols provided on the product that influenced your decision about how the product's construction is impacting the environment?*
- o Slide 2: Invite students to reflect on their concept maps and discussions with their peers. Ask them to discuss their thoughts on what makes a product environmentally friendly and aligns with environmental sustainability; generate a class list after asking students: *What characteristics did you observe that suggested a product is environmentally friendly and does not hinder environmental sustainability?*
  - ▪ *Note: While students share about their concept maps, ask them to think about whether or not they observed any elements of greenwashing given the products they analyzed (based on Unit 2: Day 2's lesson).*

Activity 2: *Company Advertising and Environmental Practices*
- Use or adapt this Google Document Graphic Organizer: https://tinyurl.com/bbc4srvc
  - o Organize students to work in small groups. Have them read/view, analyze, and discuss each text.
    - ▪ Text 1: Dawn-Dish. (2011). Dawn Dish Soap Digital Advertisement. https://dawn-dish.com/en-us
    - ▪ Text 2: Dawn Dish Soap. (2020, April 6). *It gets better at Dawn* [Video]. YouTube. https://www.youtube.com/watch?v=yn0YI1bLtBs
    - ▪ Text 3: Dawn. (2021). *Dawn dish soap ingredients.* Dash-Dish. https://dawn-dish.com/en-us/how-to/what-dawn-is-made-of-ingredients
    - ▪ Text 4: Dawn. (2021). *Saving wildlife.* Dawn-Dish. https://dawn-dish.com/en-us/dawn-saves-wildlife
    - ▪ *Extension*: Have student groups select (or assign) **one** of the texts from the graphic organizer (above). Have them analyze their text using the six CML Guiding Questions (see Unit 2: Day 1 – Activity 2 for Guiding Question PDFs). Then number-off students (1, 2, 3, 4) within their small groups. Using the jigsaw strategy (see *Cult of Pedagogy*: https://www.cultofpedagogy.com/search/Jigsaw+method/), students should form new groups according to their number and then teach their peers about the text they analyzed using the CML Guiding

Questions. Students share their expert knowledge with their peers in other groups. Conclude the lesson by having students debrief their learning process as a whole class.

Activity 3: *Environmental Practices Student Media Application*
- In partners, have students co-create a photo collage, digital (StoryBoardThat.com) or print-based comic strip to explain environmentally sustainable practices, environmental-friendly practices, and eco-labels.

Lesson Closer: *Student Media Production Share-Out*
- Students share their media creations from Activity 3 with the class and discuss their processes behind the choices they made and what their visuals/texts represent.

Day 4: *Ecowriting Alternative Media Project Introduction*

Lesson Opener: *Revisit K-W-L (What do you want to learn . ?)*
- Use or adapt this Google Presentation: https://tinyurl.com/ywzdesxv
  - Slide 1: Invite students to revisit "W" from their K-W-L charts on Day 2. Have students reflect on topics that are still of interest about greenwashing. Ask students: *What do you want to know and explore about greenwashing and its impact on the environment, consumers, and our daily lives?*
  - Slide 2: Encourage students to share any new responses from their K-W-L chart. Revisit the previous class-generated list for recommendations on new areas of exploration. Guide students to form small groups, based on their topics of interest.

Activity 1: *Project Overview*
- Use or adapt this Google Presentation: https://tinyurl.com/ynmbujxk
  - Slide 1: Review definitions of ecowriting and countermedia (alternative media) with students.
  - Slide 2: Review project guiding questions and learning outcomes.
  - Slide 3: Review project goals and the project process.

Activity 2: *Designing inquiry-based questions*
- Use or adapt this Google Presentation: https://tinyurl.com/2bzdja4j
  - Slide 1: Review the inquiry-based learning framework with students. Explain that inquiry-based learning is an ongoing process

driven by student-created open-ended questions that require critical thinking.
- Slide 2: Show an example of an inquiry-based learning project related to greenwashing to help contextualize what is an open-ended question. Ask students to watch: Our Changing Climate. (2018, November 23). *Is fast fashion destroying our environment?* [Video]. YouTube. https://www.youtube.com/watch?v=YOA0D 0i5-fA. Encourage students to write notes about what techniques the creators of this video use to respond to their questions, walk readers/viewers through their research topic, and their process of storytelling.
- Slide 3: Debrief and discuss students' observations from the video and generate a class list.
- Slide 4: Invite students to design their own inquiry-based questions related to their project topic. Guide students to complete the prompted steps on the slide.
  - Note: *Look at additional teacher resources provided in the Unit 2 instructional resource document underneath the inquiry-based learning subsection for more information.*
- Slide 5: Ask each team to share their inquiry-based learning question. Encourage students to share feedback with each other by providing guiding questions, such as: *Is the question open-ended (not answerable with a yes or no response)? Does the question spark investigation about a topic related to greenwashing? If so, how? Does the question endeavor toward environmental accountability, sustainability, and/or eco-friendly practices? If so, how? Are there any resources that you recommend?*

Lesson Closer: *Team Meetings*
- Facilitate student engagement in a team meeting to discuss the feedback they received. Encourage them to make decisions about modifications to their inquiry-based questions.

Day 5: Collecting Source Materials

Lesson Opener: *Text Types*
- Use or adapt this Google Presentation: https://tinyurl.com/5b2k4y4h
  - Slide 1: Invite students to engage in a Quick Write using guiding questions: *What kinds of texts would you like to explore to investigate*

your inquiry-based questions? Where can you find these resources? Encourage students to share their ideas with the class.
- o Slides 2–5: Review elements of critical media literacy with students.
- o Slide 6: Have students add to their lists, considering resources beyond print-based texts. Help students think about collecting multiple types of resources from different perspectives.

Activity 1: *Critical Media Analysis*
- Use or adapt this Google Presentation: https://tinyurl.com/y6sucuj2
  - o Slide 1: Have students select any text related to their inquiry-based question. Guide students to first read/view the text for comprehension and to take notes about key points, observations, or quotes that relate to their project topic.
  - o Slide 2: Encourage students to use the six CML Guiding Questions to analyze the text.
  - o Slide 3: Ask students to share about the text they examined and their responses to the six CML Guiding Questions with their teammates.

Activity 2: *Critical Media Analysis Process Debrief*
- Have students describe what they learned when engaging with the text. Ask: *What guiding questions did you find easier to respond to? Why? What guiding questions were more challenging to answer? Why?* Help students think about additional resources they would like to locate and additional areas of exploration for their research topics.
  - o *Note: Check for students' understanding of each of the six CML Guiding Questions. Provide clarification and review any of the core concepts and guiding questions (as needed).*

Lesson Closer: *Team Meetings*
- Have student teams meet to brainstorm additional resources and areas of exploration to respond to their inquiry-based question and research topic. Assist students as they delegate roles and consider their next steps for gathering resources.
  - o *Note: Students should continue to collect and analyze texts related to their inquiry topic. Encourage them to keep using the six CML Guiding Questions to analyze their texts. Additional teacher resources to assist students with analyzing sources are provided in the Unit 2 instructional resource document.*

Day 6: Synthesize Your Research

Lesson Opener: *Golden Inquiry*
- Use or adapt this Google Presentation: https://tinyurl.com/bhw46ca6
    - Slide 1: Have project groups meet and review their research notes and consider a meaningful excerpt or direct quote they found in one of their texts.
    - Slide 2–6: Using Google Slides (or sheets of paper), have a student from each team add a quote to a single Google Slide (or piece of paper). Then, encourage students to respond to the quote provided by another team member.
    - Slide 7: Provide time for each team to read each other's responses and to debrief their process with each group. Offer guiding questions, such as: *What did you notice about the information your peers' shared? What did you discover differently about your inquiry topic? What else would you like to explore?*
    - Slide 8: Facilitate a whole group reflection by having student groups describe their process and what they learned from one another. Invite students to provide feedback to other teams about any recommended texts or areas of research connected with their topics.

Activity 1: *Greenwashing Inquiry Topic Elevator Speech*
- Use or adapt this Google Presentation: https://tinyurl.com/m7zv38nh
    - Slide 1: Ask students how they would describe and share an example of an elevator speech. Ensure that students understand that an elevator speech is a brief statement (as if you are in an elevator and only have the time between floors to talk) that describes or informs about a topic with concise claims, evidence, and goals. Invite students to reflect upon discussions about advertising techniques from Lessons 1–2. Explain to students that they will practice synthesizing the research they have collected about their greenwashing inquiry topic by creating elevator speeches.
    - Slide 2: Review approaches and goals to drafting elevator speeches about students' greenwashing inquiry sub-topics.
    - Slides 3–8: Have students meet in their groups to draft their elevator speech using slides 3–8 that match their group number. After students begin drafting their elevator speeches, pick a group

member to present and practice the speech before sharing with the class.
- Slide 9: Invite student groups to share their elevator speeches. Other students should actively listen and take notes about what they observe and the information that is shared. Pause after each group shares their elevator speech to provide time for students to anonymously engage in the *Two Stars and One Wish* feedback protocol to provide two affirming comments and one constructive response to each other (see free printable template: https://tinyurl.com/2p93thuw).
- *Extension:* To add an additional layer of critical media production, invite students to record their elevator speeches using a Voice Memo recording app on a mobile device, tablet, desktop/laptop, or this free voice recording tool: https://vocaroo.com/. Students could record their original elevator speeches and/or revise their drafted speeches to incorporate feedback from their peers after collecting further resources and engaging in revision.

Lesson Closer: *Team Debrief*
- Have research teams meet and discuss feedback on their elevator speeches. Encourage them to discuss additional research or modifications they would like to make when describing their inquiry topics, incorporating critical stances, etc. Students should make decisions about whether they need to collect and analyze additional texts related to their inquiry topic. Remind students to refer to the six CML Guiding Questions as they gather resources to create their alternative media project on greenwashing.

Day 7: Countermedia Project Development

Lesson Opener: *Revisiting Countermedia*
- Use or adapt this Google Presentation: https://tinyurl.com/53xmvbp7
  - Slide 1: Activate students' prior learning about countermedia during Lesson 1. Ask students to define and describe countermedia as a class, referring to the images provided on the slide. Co-develop a one-sentence definition to describe countermedia as a class.
  - Slide 2: Add the class definition of countermedia on slide 2. Compare and contrast both definitions; include modifications from students.

- Slides 3–7: Showcase various examples of countermedia products in different mediums (e.g., video) and genres (style of communication). Throughout each example, instruct students to take notes of images, narration, dialogue, etc. that were utilized to convey a message. Encourage students to consider the CML Guiding Questions. Discuss this analysis after each countermedia example.
  - Slide 3: Example 1 (Medium: Video). OwlSwap Sustainability Initiative. (2020, November 30). *What is greenwashing? Part 1* [Video]. YouTube. https://www.youtube.com/watch?v=5_SGWaAcyyw
  - Slide 4: Example 2 (Medium: TikTok). BBC My World. (2021, October 19). *Greenwashing: What you need to know – BBC My World #shorts* [Video]. YouTube. https://www.youtube.com/watch?v=QphGrZe4b8w&t=6s
  - Slide 5: Example 3 (Medium: Webpage). Lily. (2021, October 19). *Is SHEIN ethical or sustainable? Deep Dive into their greenwashing.* Imperfect Idealist. https://imperfectidealist.com/is-shein-ethical-or-sustainable/
  - Slide 6: Example 4 (Medium: Instagram). Lily [@imperfectidealist]. (2022, January 24). Stop the #WorstWageTheft in fashion [Text/Image Thread]. Instagram. https://www.instagram.com/p/CZIADkKp9ol/?utm_medium=share_sheet
  - Slide 7: Example 5 (Medium: Infographic, Narrative Ecowriting Blogpost) Christie [@imperfectidealist]. (2020). Fast fashion: A no-regrets break-up. *Cedar + Surf.* https://cedarandsurf.com/blog/fast-fashion-a-no-regrets-breakup
  - Slide 8: Share examples of different genres, print and non-print mediums, and digital tools. Invite students to share with the class any additional ideas, mediums, and digital tools.
  - *Note: Remind students that they will create countermedia that applies ecowriting to express their research about their greenwashing inquiry topic. Check for students' understanding about the goals for this project discussed in Lesson 4. Ask students to consider which medium and genre-format they would like to use to communicate their learning.*

Activity 1: *Team Discussion*
- Have students meet in their inquiry teams. During their meetings students should make progress on their countermedia product development:

Ecowriting Unit 2

- Select medium (print, non-print, mixture of both), genre(s), digital tools.
- Remind students that, depending on their medium and genre preferences, they should feel free to edit/revise their elevator speeches (Lesson 6) to be suitable for the medium/genre they select.
- Designate roles (if needed): Make sure that everyone has a say in which role they would like to take on during the production process to ensure that everyone's talents, preferences, and contributions are valued. *For example, researchers (collect/synthesize additional print and non-print research, including sounds, images, or video clips, etc.), graphic designer/artist (finds or creates images), ecowriters (adapt elevator speech and/or draft new writing to match the genre-style approach of storytelling), etc.*
- Storyboard: Have students create a storyboard as a group to plan out their ecowriting countermedia product. Use paper, online document tool (https://docs.google.com), visualization tool (https://jamboard.google.com), etc. After completion, students will share with the class.

Activity 2: *Storyboard Peer Feedback*
- Pass out sticky notes, index cards, or half sheets of paper to each student. Students should have enough materials to provide feedback for each team (e.g., if there are six inquiry teams, every student should receive five sticky notes). Ask the students to label their materials with the group number at the top and to draw a t-chart with columns labeled *I like* and *I wonder*. Provide time after each group shares their storyboard/plans for classmates to provide anonymous feedback.

Lesson Closer: *Team Debrief*
- Provide time for each inquiry group to review their peer feedback and to set subsequent goals for their ecowriting countermedia project development.

Day 8: Countermedia Project Development Session 2

Lesson Opener: *Project Check-in*
- Check in with each inquiry group about their progress. Provide feedback, answer questions, and help guide students as they develop their ecowriting countermedia products.

- *Note: Depending on the students' progress, adjust the number of instructional sessions you would like to provide as project workdays. If students are contributing to this production outside of class, the time needed for critical media production will vary. Number of project work sessions may vary for age-levels, technology access, and learning styles.*

Activity 1: *Inquiry Team Workshop*
- Inquiry teams continue their ecowriting countermedia project productions. Circulate around the room and check in with groups (as needed), observe what roles they are engaging in within their groups, and offer support and guidance. Toward the end of the session, remind students to set goals and review team roles to ensure action steps are understood.

Lesson Closer: *Team Updates*
- Each team shares a brief update about their progress and next steps. Celebrate students' accomplishments as a class and answer any remaining questions.

Day 9: Dress Rehearsal

Lesson Opener: *Project Showcase Format*
- Check in with each team to see how they are feeling about the upcoming project showcase. Field any remaining questions. Explain project showcase format (e.g., presentation order, attire, technology availability, where to display non-print products, etc.). Ask students for feedback about the project showcase format and make any adjustments.
  - *Note: Depending on how the students will share their work will impact the design of the dress rehearsal. For example: Students presenting in an auditorium would need to know where they should sit when they are not presenting, the order they will walk up to the presentation area, technology available, or where/how to display what they have created. For an online format, students would need to practice sharing screen or making sure their instructor has their materials in advance, etc.*

Activity 1: *Team Meeting*
- Have student groups practice and determine how they will present (e.g., how they will set-up a video clip, who is speaking when, who

is facilitating technology components or setting up physical projects, etc.).
- Note: If needed, before students begin practicing how they will deliver their presentation, show the video by Practical Psychology. Ask students about their observations and other ideas for delivering an engaging presentation. See: Practical Psychology. (2017, January 16). *How to give a great presentation – 7 presentation skills and tips to leave an impression* [Video]. YouTube. https://www.youtube.com/watch?v=MnIPpUiTcRc

Activity 2: *Rehearsal*
- Student groups present their projects to the class. Audience provides feedback to each group (orally). Ask the group receiving feedback to take notes about audience recommendations.

Lesson Closer: *Finalization*
- Have students meet in teams to review the feedback given by the audience. Student inquiry groups should make final decisions about any remaining adjustments to improve their products and presentation approaches. Review project showcase expectations; address student questions.

Day 10: *Project Showcase*

Lesson Opening: *Project Introduction*
- Offer opening remarks and welcome students, families, community members, and guests. Provide context to the audience about environmental sustainability, environmental ethics, and greenwashing. Describe the scope of the project the students have been engaging in during this unit. Celebrate the accomplishments of the students and thank the audience for attending.

Activity 1: *Project Showcase*
- Students share what they have learned with their family, friends, and community.
  - Note: Depending on showcase format, invite the audience to ask questions, engage in a dialogue with presenters after each presentation, or at the end of the showcase.

Lesson Closer: *Closing Remarks and Gratitude*
- Give closing remarks and thank attendees. Congratulate students on their meaningful collaboration and learning processes throughout

the project. Encourage audience members to reflect on students' countermedia products and calls-to-action to take individual, local, and global actions for climate justice.

Additional Ecowriting Extension Lesson: Celebrating Environmentally Friendly Practices

Lesson Opening: *Reflection*
- Explain Socratic seminars to students using video by Let's TEACH. Clarify questions about how to engage in a Socratic seminar.
  - Let's TEACH. (2020, September 8). *Instructional Strategy – Socratic Seminar* [Video]. YouTube. https://www.youtube.com/watch?v=SW-WQk-UnUg.

Activity 1: *Class Reflection*
- Construct alternate classroom set-up for better conversation flow (e.g., chairs in a circle). Begin the Socratic seminar to help facilitate student dialogue by providing a couple of guiding questions, such as: *What are key takeaways from this unit? What actions can we take to support positive environmental practices?*

Lesson Closer: *Celebrating Envrionmentally Friendly Actions*
- To continue students' ecowriting, they will produce a letter to the company or organization of their choice. Each student should select a company or organization and look at their webpage to learn more about them. Ask students to browse the EcoLabels website: Big Room, Inc. (2022). *EcoLabel Index.* https://www.ecolabelindex.com/ecolabels/.
- Express support to companies and organizations that are helping the environment and do not participate in greenwashing by using the letter generator: ReadWriteThink. (2022). *Letter Generator.* NCTE. https://www.readwritethink.org/classroom-resources/student-interactives/letter-generator#overview
- Send letters to companies and organizations. Encourage students to continue their ecowriting journeys and encourage them to continue taking actions to help create a more environmentally sustainable world.

30. Ecowriting Unit 2: Greenwashing – Instructional Resources Guide

SYDNEY RICHMOND AND ANDREA GAMBINO

*Digital unit version available at: https://tinyurl.com/bdhrvkwm

Figure 30.1: Scan this QR code for access to the digital version of this unit.

Instructions: To access all Google Presentations and resources described in Ecowriting Unit 2: Greenwashing, follow the guide below. External links in lesson plan activities are embedded in each Google Presentation (slides and/ or presenter notes). Adapt any of the Google Presentations for your own use with students to be developmentally appropriate and culturally relevant for your classroom context. To modify the Google Presentations for your classroom, click *file, make a copy, and save them to your local Google Drive.* Remember to edit the share settings so that students can access the materials.

Day 1: Introduction to Critical Media Literacy and Commercial Media/Advertising

- Lesson Opener: *Logo/Brand Alphabet* https://tinyurl.com/62rvjr3k
- Activity 1: *McDonalds' Advertisement Analysis* https://tinyurl.com/4vb97pap
- Activity 2: *What is Critical Media Literacy? How can we use the CML questions?* https://tinyurl.com/5wzzkx4a
- Critical Media Literacy Framework: Conceptual Understandings and Guiding Questions (Kellner & Share, 2019, p. 8) with PDFs available in:
  - English: https://tinyurl.com/4v5ndatx
  - Spanish: https://tinyurl.com/2fdaz8up
  - Mandarin: https://tinyurl.com/mpfveh85
  - Portuguese: https://tinyurl.com/mwrkf9sz
- Activity 3: *Dominant and Countermedia* https://tinyurl.com/a8v752ea

Day 2: Introduction to Greenwashing

- Lesson Opener: *Greenwashing K-W-L Chart ("I know ..")* https://tinyurl.com/nfvakdrf
- Activity 1: *"A Fiji Water Story" and Greenwashing Techniques* https://tinyurl.com/un8956uu
- Activity 2: *Greenwashing K-W-L Chart ("I want to know ..")* https://tinyurl.com/ddbhus25
- Activity 3: *Greenwashing Reading Groups* https://tinyurl.com/24yeyze9
- Lesson Closer: *Greenwashing K-W-L Chart ("I learned ..")* https://tinyurl.com/7y5prmkk

Day 3: Environmentally Friendly Practices, Environmental Sustainability, and Eco-labels

- Lesson Opener: *Defining Environmentally Friendly, Environmental Sustainability, and Eco-labels* https://tinyurl.com/dm8reh3t
- Activity 1: *Conscious Consumption Product Mapping* https://tinyurl.com/vphh3vtk
- Activity 2: *Company Advertising and Environmental Practices* https://tinyurl.com/bbc4srvc

Day 4: Ecowriting Alternative Media Project Introduction

- Lesson Opener: *Revisit K-W-L (What do you want to learn ..?)* https://tinyurl.com/ywzdesxv
- Activity 1: *Project Overview* https://tinyurl.com/ynmbujxk
- Activity 2: *Designing inquiry-based questions* https://tinyurl.com/2bzdja4j

Day 5: Collecting Source Materials

- Lesson Opener: *Text Types* https://tinyurl.com/5b2k4y4h
- Activity 1: *Critical Media Analysis* https://tinyurl.com/y6sucuj2

Day 6: Synthesize Your Research

- Lesson Opener: *Golden Inquiry* https://tinyurl.com/bhw46ca6
- Activity 1: *Greenwashing Inquiry Topic Elevator Speech* https://tinyurl.com/m7zv38nh

Day 7: Countermedia Project Development

- Lesson Opener: *Revisiting Countermedia* https://tinyurl.com/53xmvbp7

Additional Curricular and Pedagogical Resources for Educators

Instructions: The following resources can aid in curriculum planning and pedagogical understandings for educators to explore approaches to critical media literacy, inquiry-based learning, and analyzing source materials.

Critical Media Literacy[1]: An inquiry-based process for analyzing and creating media by interrogating the relationships between power and knowledge. It questions representations of class, gender, race, sexuality and other forms of identity and challenges unjust media messages. CML celebrates positive representations and beneficial aspects of media while challenging problems and negative consequences, recognizing media are never neutral.

---

[1] This definition is based on the definition of Critical Media Literacy generated by the 2021 Critical Media Literacy Conference of the Americas Steering Committee, see: https://tinyurl.com/t29f89u8.

- Critical Media Project. (2022). https://criticalmediaproject.org/
- Critical Media Literacy Conference of the Americas (2021) videos and resources for researchers, educators, and public intellectuals are hosted by the Critical Media Project and available here: https://criticalmediaproject.org/conferences/
- Kellner, D., & Share, J. (2019). *The critical media literacy guide: Engaging media and transforming education.* Brill/Sense Publishers. 10.13140/RG.2.2.32448.79360. https://brill.com/view/title/55281
- Critical Media Literacy Framework Conceptual Understanding & Guiding Questions (Kellner & Share, 2019, p. 8) is available for free in multiple languages:
  - English: https://tinyurl.com/4v5ndatx
  - Spanish: https://tinyurl.com/2fdaz8up
  - Mandarin: https://tinyurl.com/mpfveh85
  - Portuguese: https://tinyurl.com/mwrkf9sz
- Morrell, E., Dueñas, R., Garcia, V., & Lopez, J. (2015). *Critical media pedagogy: Teaching for achievement in city schools.* Teachers College Press. https://www.tcpress.com/critical-media-pedagogy-9780807754382
- Project Censored & The Media Revolution Collective. (2023). *The media and me: A guide to critical media literacy for young people.* The Censored Press. https://www.project-censored.org/shop/p/the-media-and-me-a-guide-to-critical-media-literacy-for-young-people
  - Free to access companion teaching guide for educators using the above book available here: https://www.projectcensored.org/wp-content/uploads/2022/12/Media-Me-Teaching-Guide-FINAL.pdf
- Share, J., Gambino, A., & Marineo, F. (Eds). (updated regularly). Critical media literacy research guide. *University of California, Los Angeles.* https://guides.library.ucla.edu/educ466

Inquiry-Based Learning: A student-centered pedagogical approach that encourages learners to ask questions, think critically, post real-world solutions, and prioritizes experiential learning guided by students' curiosity.

- Education Development Center, Inc. – YouthLearn. (2016). Inquiry-based learning: An approach to educating and inspiring kids. http://youthlearn.org/wp-content/uploads/Inquiry_Based_Learning.pdf

- Edutopia. (2015, August 24). *Inquiry-based learning: Developing student-driven questions* [Video]. YouTube. https://www.youtube.com/watch?v=OdYev6MXTOA
- Heick, T. (2018, February 26). Many, many examples of essential questions. *TeachThought.* https://www.teachthought.com/pedagogy/examples-of-essential-questions/ (*Note: In this resource, the author refers to question types as *essential questions*; however, they could also be referred to as inquiry-based questions).
- Go Pangea. (n.d.). Essential question stems. https://www.gopangea.org/essential-question-stems-for-inquiry-based-learning
- Spencer, J. (2017, December 5). *What is inquiry-based learning?* [Video]. YouTube. https://www.youtube.com/watch?v=QlwkerwaV2E
- Spires, H. A., Kerkhoff, S. N., & Paul, C. M. (2020). *Read, write, inquire: Disciplinary literacy in grades 6–12.* Teachers College Press. https://www.tcpress.com/read-write-inquire-9780807763339
- TeachThought. (2017, July 29). 28 critical thinking quiet stems for any content area. https://www.teachthought.com/critical-thinking/critical-thinking-question-stems-content-area/

Analyzing Sources: Providing scaffolds to analyze, question, and respond to source materials.

- Benedictine University. (2021, January 19). *Evaluating Sources: The CRAAP Test – Currency, relevance, authority, accuracy, and purpose.* Research Guides – Evaluating Sources. https://researchguides.ben.edu/source-evaluation
- Common Sense Media. (2021, October). *Grade 6 – Finding Credible News: How do we find credible information on the internet?* https://www.commonsense.org/education/digital-citizenship/lesson/finding-credible-news
- McKillop Library. (2022). *Finding credible news.* McKillop Library Research Guides. *Salva Regina University.* https://salve.libguides.com/credibleNews/home
- News Literacy Project. (2022). https://newslit.org/
- Project Censored. (2022). *How to find, evaluate, and summarize independent news stories.* https://www.projectcensored.org/how-to-find-evaluate-and-summarize-validated-independent-news-stories-vins-project-censored/

- Randolph Community College Library. *Scholarly vs. popular articles.* https://s3.amazonaws.com/libapps/accounts/59233/images/Scholarly_VS_PopularPSY.jpg
- Share, J., Gambino, A., & Marineo, F. (eds). (updated regularly). *Journalism and news.* Critical media literacy research guide. *University of California, Los Angeles.* https://guides.library.ucla.edu/c.php?g=1108715&p=8086728

31. Recommended Resources

HTTPS://TINYURL.COM/YR9ENDHN

Figure 31.1: Scan this QR code for access to the digital version of this recommended resources document.

Websites
- This UCLA *Critical Media Literacy Library Research Guide* provides links to articles, videos, lesson plans, podcasts and an assortment of free resources related to critical thinking about media, information, and technology. https://guides.library.ucla.edu/educ466
- This wiki contains many resources for teaching about the climate crisis. It is a companion to the book, *Teaching Climate Change to Adolescents*, listed below: https://tinyurl.com/2dus9pkh
- *Action for the Climate Emergency* (ACE) is a nonprofit organization committed to educating young people about the science of climate change and empowering them to take action. They provide lots of material from videos to lesson plans. https://acespace.org/
- *Climate Lit i*s a peer-reviewed open access online database fostering climate literacy for young people through literature, film, and

various media. It includes a glossary, resources, and reviews. https://www.climatelit.org/

Children's Picture Books
- *We Are Water Protectors* – Written by Carole Lindstrom, Illustrated by Michaela Goade – a young girl from the Ojibwe nation fights to protect her water against the Dakota Access Pipeline.
- *Weeds Find a Way* – Written by Cindy Jenson-Elliot, Illustrated by Carolyn Fisher – a heroic story about the resilience of weeds to always find a way to thrive.
- *Shi-shi-etko* – Written by Nicola I. Campbell, Illustrated by Kim LaFave – her last days before being sent to Indian Residential School, a young Indigenous girl memorizes all her connections with the natural world so that she will never forget them.
- *Sky Sisters* – Written by Jan Bourdeau Waboose, Illustrated by Brian Deines – a night journey in the frozen countryside for two Ojibway sisters searching for the SkySpirits.
- *My Wounded Island* – Written by Jacques Pasquet, Illustrated by Marion Arbona – a story about the Iñupiat people who live on an island near the Arctic Circle being consumed by rising sea levels caused by an evil creature symbolic of global warming.
- *A River Ran Wild* – Written and Illustrated by Lynne Cherry – a true story about the history of the Nashua river from Indigenous sustainable relationships to colonial exploitation and contamination, and ultimately to the struggle to clean it up.
- *Wangari's Trees of Peace* – Written and Illustrated by Jeanette Winter – a true story from Africa about Wangari Maathai, Nobel Prize recipient who started the Green Belt Movement in Kenya.
- *The Woman Who Outshone the Sun* – From a poem by Alejandro Cruz Martínez, Illustrated by Fernando Olivera – a Zapotec Indian legend from Oaxaca, Mexico about living in harmony with nature and treating everyone with respect.
- *The Keeper of Wild Words* – Written by Brooke Smith, Illustrated by Madeline Kloepper – a child and her grandmother work to save "wild words" about nature that are disappearing because of their lack of use.
- *My Friend Earth* – Written by Patricia MacLachlan, Illustrated by Francesca Sanna – a journey across the globe and through the seasons appreciating the natural world.

Books for Teachers about Teaching Environmental Justice

- *Teaching Climate Change to Adolescents: Reading, Writing, and Making a Difference,* by Richard Beach, Jeff Share, and Allen Webb (2017), Routledge and NCTE. Written for English teachers, this book provides theory and practical lessons for teaching all students about climate change.
- *A People's Curriculum for the Earth: Teaching About the Environmental Crisis* edited by Bill Bigelow and Tim Swinehart. Rethinking Schools. Articles and lesson plans are brought together in this teacher-friendly book that provides all the materials for role plays, simulations, stories, and poems for teaching environmental justice.
- *Education in Times of Environmental Crises: Teaching Children to be Agents of Change* edited by Ken Winograd (2016) – this collection of essays offers educators guidance from Nel Noddings on the importance of biophilia to lessons from Lynne Cherry about a film project on the environment.
- *Teaching Truly: A Curriculum to Indigenize Mainstream Education* by Four Arrows (Don Trent Jacobs) (2013) – Through incorporating history, culture, and worldviews, Four Arrows asserts the value of Indigenous worldviews and pedagogy over mainstream colonial educational practices most common in Western societies.
- *EcoJustice Education: Toward Diverse, Democratic, and Sustainable Communities* (2nd ed.) by Rebecca A. Martusewicz, Jeff Edmundson, and John Lupinacci (2015) – this book provides a critical approach to environmental education, with in-depth essays addressing the cultural foundations of the ecological crisis, enclosing of the commons, focus on race, class, gender, and Indigenous perspectives.

Books About Environmental Justice
- *As Long as Grass Grows: The Indigenous Fight for Environmental Justice, From Colonization to Standing Rock* by Dina Gilia-Whitaker (2019) – a historical overview of the devastating effects of settler colonialism on Native Americans and the centuries of Indigenous struggle provides a powerful look at environmental justice from an Indigenous perspective.
- *Black Nature: Four Centuries of African American Nature Poetry* edited by Camille T. Dungy (2009) – filled with African American poets that focus their writing on nature.
- *Spiritual Ecology: The Cries of the Earth* edited by Llewellyn Vaughan-Lee (2016) – essays and poems written by several people about the

ecological crisis we are currently facing and how it is connected to "our forgetfulness of the sacred nature of creation."
- *Braiding Sweetgrass: Indigenous wisdom, scientific knowledge, and the teachings of plants* by Robin Wall Kimmerer (2013) – as an Indigenous scientist, Kimmerer explores relationships between plants and animals and the need for reciprocity, respect, and responsibility.
- *Climate Justice: Hope, Resilience, and the Fight for a Sustainable Future* by Mary Robinson with Caitríona Palmer (2019) – The former president of Ireland and UN Special Envoy on Climate Change unites the battle against climate change with the struggle for human rights through stories about inspirational women leading the way.

Movies
- *Awake: A Dream from Standing Rock* – a documentary that "captures the story of Native-led defiance that forever changed the fight for clean water, our environment and the future of our planet" (https://awakethefilm.org).
- *Fantastic Fungi* – goes into the deep, wonderful powers that fungi offer our world such as its medicinal properties, communication, regeneration, sustainability, and more. https://fantasticfungi.com/
- *Public Trust* – the fight for preservation of the 640 million acres of public lands in the United States. https://www.patagonia.com/films/public-trust/
- *The Condor and the Eagle* – Indigenous leaders from Canada to the Amazon connect their struggles and activism for climate justice. (https://filmsfortheplanet.com/the-condor-the-eagle/
- *This Changes Everything* – inspired by Naomi Klein's 2014 book of the same name, this documentary about the climate crisis aims to empower viewers into action. https://thischangeseverything.org/the-documentary/
- *Virunga* – a true story about people fighting to protect the Virunga National Park in the Congo from groups and companies trying to exploit and destroy the land for its natural resources. https://filmsfortheearth.org/en/film/virunga/

Short Videos
- *The Honorable Harvest* by Robin Wall Kimmerer, https://www.youtube.com/watch?v=cEm7gbIax0o&t=9s – this 3:30 minute video addresses the importance and sacredness of plants, and how we must listen and learn from them as our relatives and teachers
- *Story of Solutions* – from the Story of Stuff Project, this 9-minute video explains the differences between living with our current

consumeristic goal of "more" versus a sustainable goal of "better." https://www.storyofstuff.org/movies/the-story-of-solutions/
- *Ecomedia* by Aaron Kierbel – based on Antonio López's work about ecomedia literacy, this 2:30 minute animated video shows the impact that a smartphone has on the planet. https://www.youtube.com/watch?v=JUbFJwqzP1Q
- *Ron Finley: A Guerrilla Gardener in South Central LA* – this 10:30 minute TED Talk by Ron Finley, focuses on his dedication to gardening vegetables in South Central Los Angeles, which is known to be a food desert. https://www.ted.com/talks/ron_finley_a_guerrilla_gardener_in_south_central_la?language=en
- *Protect Mauna Kea – Damian "Jr Gong" Marley* – Damian Marley visits Mauna Kea to support the efforts Native Hawaiians have put forth to protect this sacred mountain against the construction of a telescope (3:40 minutes). https://www.youtube.com/watch?v=gQVWNqkb2H8
- *Indigenous Worldview can Preserve our Existence* – Narrated by Four Arrows, this short video highlights the importance of recognizing our interconnections with the natural world and the potential of embracing Indigenous worldviews, "we are all related" (2:52 minutes) https://www.youtube.com/watch?v=QkQTeVmHn7M

Songs
- *Black Snakes* by Prolific the Rapper x A Tribe Called Red https://www.youtube.com/watch?v=QdeHUrL1FEM – a song is about the efforts of the water protectors during their fight against the construction of the Dakota Access Pipeline
- *New Beginning* by Raskahuele (Spanish) https://www.youtube.com/watch?v=adk84FsJa1c – sings on the destruction and oppression the world is currently facing due to our lack of respect for Mother Earth.
- *My Hawai'i* by The Green https://www.youtube.com/watch?v=oOVG46AliXc – a song dedicated to Hawai'i and the importance of protecting the land and its people.
- *Symphony of Science: Our Biggest Challenge* – a musical investigation into the human causes and effects of the climate crisis, including Bill Nye, David Attenborough, Richard Alley and Isaac Asimov. https://www.youtube.com/watch?v=HHP9Rh-ooh0

Poetry
- *Earthrise* by Amanda Gorman (2018) – from the 1968 *Earthrise* photo of planet Earth to our current environmental challenges,

Amanda Gorman encourages everyone to rise up and protect our delicate home. https://www.youtube.com/watch?v=xwOvBv8RLmo
- *Hozhó* by Lyla June (2012) – Lyla June shares her grandmother's teachings about the Diné Bizaad (Navajo language) word Hozhó, as the sun rises over the landscape. https://www.youtube.com/watch?v=PZzPWKJvu7I
- *Agua, Agüita / Water, Little Water* by Jorge Argueta (2017) – a trilingual book written in verse about the life-giving force that is water and its many forms it takes from raindrops to the oceans and rivers. https://www.youtube.com/watch?v=_9bRdZytQBw
- *MAN vs EARTH* by Prince Ea (2015) – spoken word and images about the problems of environmental degradation and the importance of coming together to battle against climate change. https://www.youtube.com/watch?v=VrzbRZn5Ed4
- *Habitat Threshold* by Craig Santos Perez (2020) – eco-poetry that explores the author's Pacific Islander heritage, the ecological situation of his homeland, and his concerns for the future. https://press.uchicago.edu/ucp/books/book/distributed/H/bo50382597.html

Podcasts
- *Change Everything*, a podcast from LEAP for people concerned about the climate, racism, and inequality to advance intersectional solutions to overlapping crises (https://theleap.org/our-work/change-everything/)
- *The Bioneers: Revolution from the Heart of Nature*, an international radio and podcast series with voices of grassroots activists who are often excluded from commercial media. (https://bioneers.org/bioneers-radio/)
- *Earthwatch*, connects people with scientists around the world through citizen science and community engagement (https://earthwatch.org/stories/podcasts)
- *Got Science?* Podcast by the Union of Concerned Scientists. This U.S. nonprofit organization was created by scientists and students at the Massachusetts Institute of Technology (MIT). (https://www.ucsusa.org/resources/all/podcast)
- *A Sustainable Mind* podcast by Marjorie Alexander that investigates environmental campaigns and movements. (https://asustainablemind.com/)
- *Climate One* podcast began in 2007 and is sponsored by *The Commonwealth Club* to explore long-term solutions to climate

disruption with an engaged public. (https://www.climateone.org/watch-and-listen/podcasts)
- *Drilled* is a true-crime podcast about climate change, hosted and reported by award-winning investigative journalist Amy Westervelt. (https://drilled.media/
- *Scene on Radio*, season five, "The Repair" contains 11 episodes with hosts John Biewen and Amy Westervelt as they explore the cultural roots of the climate crisis. (https://www.sceneonradio.org/the-repair/)

Notes on Contributors

SARAWI ANDRANGO is a goldsmith, embroiderer, farmer, cultural manager, artistic producer, poet, writer and lyrical composer of the Kichwa nation, Kayambi people. She studied Law, Political Science and Public Management. She has participated in international poetry readings and book fairs in Venezuela, Peru, Bolivia, Mexico and Ecuador. She has published three poetry books and a book of Andean stories.

JUSTIN C. M. BROWN is a Los Angeles-based artist, musician, and writer studying Sociology at UCLA. He cordially invites you to browse his art at https://JCMBmade.com/.

ANTONIA BURGARD earned a B.A. in English, Mathematics and Education from University of Münster and is currently pursuing a MEd. She is passionate about Literature and educational Psychology and has been active in youth work for several years, mentoring young people and encouraging them to think creatively and critically.

DENISE CHAPMAN, Ed.D. is a storyteller, spoken word poet, and teacher educator who lectures in children's literature, early literacy, and inclusive children's media at Monash University, Australia. Her research explores the use of storytelling, poetry, children's literature, and digital images as counternarrative windows for social change.

RUCHA DESHPANDE is a recent graduate from UCLA, earning a B.S. in Neuroscience and a minor in Education Studies. She is currently a medical student at the USC Keck School of Medicine and plans to become a physician.

SARA FERNANDEZ is a former student at UCLA studying Chicanx/Central American studies and education. She also is a current model aspiring to break boundaries in the realm of fashion. Sara strives to keep revolutionizing the modeling industry and also to become a future educator within her community of North East Los Angeles.

ANDREA GAMBINO is a former English/history teacher. She earned her B.S. in Middle Grades Education and M.Ed. in New Literacies and Global Learning (North Carolina State University) as well her Ph.D. in Education (UCLA). Her research centers teachers' embodiments and practices of critical media literacy. https://www.andreagambinophd.com/

MELISSA GREENE-BLYE, Ph.D. examines journalistic representations and negotiations of American Indian identity, issues, and individuals past and present. She worked as a broadcast journalist for 20 years before joining the University of Kansas School of Journalism in 2020, where she enjoys educating the next generation of journalists. Melissa is an enrolled citizen of the Miami Nation.

NICOLE HALL is a first-generation college graduate from UCLA with a B.A. in American Literature and Culture and a minor in Education Studies. Shen is working to receive a teaching credential and become a social and environmental justice educator.

PEACHES HASH is a former high school English teacher and current lecturer of Rhetoric and Writing Studies in Appalachian State University's Department of English. She holds degrees in Educational Leadership, Curriculum and Instruction, and English. Her current research involves expressive arts, practitioner action research, and writing.

LEA IRINA HEUING is studying for the Master of Education at the University of Münster, Germany, to become a teacher for vocational schools. She holds a Bachelor of Arts in English, health sciences, and education.

CINDY JENSON-ELLIOTT, M.A., writes and teaches science and language arts in southern California. She is the author of 17 nonfiction books for children. She has led students from preschool to adult toward a love of nature in schools, universities, museums, gardens, and outdoor education programs.

JANA KRANZ holds a Bachelor of Arts in English, Dutch and education and is currently studying for a Master of Education at the University of Münster,

Germany to become a secondary school teacher. She is also currently working as a nursery school teacher in a German kindergarten.

GAVIN LAMB is a sociocultural and applied linguist exploring the intersection of language, culture and the natural environment. He earned his Ph.D. in applied language and communication studies from the University of Hawai'i at Manoa, specializing in intercultural and environmental communication. His research draws on a combination of ethnography and discourse analysis to explore how language and culture shape human relationships and interactions with animals and nature.

KATHY LIZAOLA earned a B.A. in Psychology with a minor in Education Studies at UCLA. She recently graduated from the MAT (Master of Arts in Teaching) program at USC, earning a single-subject teaching credential in English.

ANTONIO LÓPEZ, Ph.D. has a research focus on bridging ecojustice with media literacy. His most recent book is *Ecomedia Literacy: Integrating Ecology into Media Education* (Routledge). Currently he is Associate Professor of Communications and Media Studies at John Cabot University in Rome, Italy. Resources and writing are available at: https://antonio-lopez.com/

KEIMORA NETTLES is an English major, education studies minor from Compton, Ca. She saw the inequalities between low-income communities and wealthy areas during her educational journey in Long Beach and coming to UCLA as a first-generation college student. Keimora plans to pursue a legal career after undergrad.

ALEJANDRO OJEDA is a current UCLA senior earning his B.A. in Education and Social Transformation and in History, with a minor in Chicanx and Central American Studies. He plans to educate future generations through a social justice framework, while highlighting the importance of art and arts programs in schools.

ESMERALDA OROZCO SÁNCHEZ is a recent UCLA graduate who earned her B.A. in Chicana and Chicano Studies and minored in Education Studies. She was recently hired as a teaching assistant at a local charter school and plans to pursue her dream of becoming an elementary school teacher.

ELMER ORTEGA is a UCLA graduate with a B.A. in Psychology and a minor in Education. He aspires to become an educator in marginalized communities with an emphasis on financial as well as critical media literacy.

JENIFER RAMOS is a recent UCLA graduate, earning a B.S. in Public Affairs with a minor in Education Studies. She plans on pursuing her dreams of working in the nonprofit sector and working collaboratively with others, using her determination to think beyond what is expected.

THERESA REDMOND is an arts-based teacher-scholar researching at the nexus of multiple fields, including media literacy, educational technology, curriculum design, and the arts. She is a former Visual Arts teacher who currently works as Associate Professor at Appalachian State University where she teaches in Media Studies and Teacher Education.

SYDNEY RICHMOND earned her B.S. at North Carolina State University studying in the Civil, Construction, and Environmental Engineering Department. Her research and industry work is centered around the NAE (National Academy of Engineering) Grand Challenges for Engineering. She aspires to receive her MS and PhD in a related field to continue combatting the challenges of our world.

VANESSA ROMERO is a UCLA graduate with a B.A. in Political Science and a minor in Education Studies. She is a paralegal working in her community at the Riverside County District Attorney's Office. Vanessa is working to pursue her dream of becoming an attorney specializing in environmental and human rights law.

NEIDA SANDOVAL-LOPEZ is a recent UCLA graduate from the Central Valley with a B.A. in Sociology. Neida dedicates her time finding creative expressive tools through art to teach marginalized communities how to navigate all social structures, keeping mental health awareness as the root, inspiration and importance of her creations.

ARBREAN SEARS is a current senior at UCLA majoring in anthropology and minoring in education. She plans on getting her teaching credential and Master's in Education and implementing environmental justice in her curricula to future students.

LUDMILLA SEMSKOW holds a Bachelor of Arts in English, German and education. She is studying for a Master of Education at the University of Münster, Germany to become a secondary school teacher. She is also currently teaching German as a second language in a German elementary school.

Notes on Contributors

JEFF SHARE has been teaching in the School of Education and Information Studies at the University of California, Los Angeles since 2007. His research and practice focuses on preparing educators to teach critical media literacy in K-12 education, for the goals of social and environmental justice. He has worked as an award-winning photojournalist and bilingual elementary school teacher in the Los Angeles Unified School District. https://jshare.wixsite.com/jeffshare

REBECCA SOLNIT is a writer, historian, activist and the author of more than 20 books on feminism, western and urban history, popular power, social change and insurrection, wandering and walking, hope and catastrophe. She writes regularly for *The Guardian* and launched the climate project Not Too Late (nottoolateclimate.com).

(ALICE) YANAN SUN is a UCLA graduate with a B.A. in History and a minor in Education Studies. She is now a graduate student at USC pursuing a Master's degree in Public Policy. Her research focuses on using statistical methods to examine the long-term impact of the COVID-19 pandemic on student learning outcomes.

BENJAMIN THOMPSON is a recent UCLA graduate with a B.A. in English and a minor in Education Studies. He is working to earn his MEd and become a high school English teacher with a focus on critical media literacy.

MANDIE TORRES is a recent UCLA graduate with a B.A. in Chicanx & Central American Studies, and a double minor in African American Studies and Gender Studies. She aspires to teach high school Ethnic Studies and push for social and environmental justice through education.

MARÍA VERÓNICA VALERIANO is a UCLA alum with a B.A. in Film and Television and a double minor in Education Studies and Spanish. She's an aspiring film editor who hopes to one day start a nonprofit that inspires the next generation of filmmakers from low-income communities.

JAZMINE VEGA LOPEZ is a Graduate student at California Lutheran University earning her M.S. in Counseling for higher education. She is a UCLA alumni with a B.A. in Gender Studies and a minor in Education Studies. Jazmine dedicates her time mentoring, assisting, and advocating for first-generation students that are from marginalized communities who would like to pursue a higher education.

GABRIELA VENEGAS is a first-generation, second year Education and Social Transformation Major from Central California. She is passionate about sustainable fashion and advocates to eradicate fast fashion companies. She aspires to be an elementary educator and help the low-income, Latinx community she grew up in.

GISELLE VILLANUEVA is a recent UCLA graduate with a B.A. in Human Biology and Society and a double minor in Education & Global Health. She is working to pursue a Master's degree in Public Health Policy. She hopes to give back to her community as a health educator.

ROSE WHITE, M.A., retired after 47 years teaching bilingual elementary school, serving in various Los Angeles Unified School District positions, and supervising teacher credentialing candidates. She attended and then trained educators in UCLA's Writing/Science/Math and Literature Projects. She also studied at the Smithsonian and trained in elementary physics/chemistry at Miami University of Ohio. She has experienced nature, art, and culture in all 50 states and 20 countries.

PHILIPPA WITZENHAUSEN is a German student at the Westfälische-Wilhelms-Universität in Münster and aspires to teach Math and English at secondary schools in the future. She focused on Digital Media Literacy as part of teaching in a seminar with Jeff Share.

YAYING WU is an undergraduate student at Northwestern University studying Sociology, Education, and Communication. Her interests in Education developed during her time as an undergraduate at UCLA. She is passionate about utilizing digital storytelling as a tool for advocating social justice and disseminating knowledge.

About the Editor

JEFF SHARE, PhD has worked for three decades researching and teaching critical media literacy and environmental justice. He was an award-winning photojournalist, bilingual elementary school teacher, and since 2007, he has been teaching in the School of Education at the University of California, Los Angeles (UCLA). With Richard Beach and Allen Webb, Share co-wrote (2017), *Teaching Climate Change to Adolescents: Reading, Writing, and Making a Difference.* In 2019, writing with Douglas Kellner, he co-authored, *The Critical Media Literacy Guide: Engaging Media & Transforming Education.* (https://jshare.wixsite.com/jeffshare)

Studies in Criticality

General Editor
Shirley R. Steinberg

Counterpoints publishes the most compelling and imaginative books being written in education today. Grounded on the theoretical advances in criticalism, feminism, and postmodernism in the last two decades of the twentieth century, Counterpoints engages the meaning of these innovations in various forms of educational expression. Committed to the proposition that theoretical literature should be accessible to a variety of audiences, the series insists that its authors avoid esoteric and jargonistic languages that transform educational scholarship into an elite discourse for the initiated. Scholarly work matters only to the degree it affects consciousness and practice at multiple sites. Counterpoints' editorial policy is based on these principles and the ability of scholars to break new ground, to open new conversations, to go where educators have never gone before.

For additional information about this series or for the submission of manuscripts, please contact:

>Shirley R. Steinberg, General Editor
>msgramsci@gmail.com

To order other books in this series, please contact our Customer Service Department:

>peterlang@presswarehouse.com (within the U.S.)
>orders@peterlang.com (outside the U.S.)

Or browse online by series:

>www.peterlang.com

www.ingramcontent.com/pod-product-compliance
Ingram Content Group UK Ltd.
Pitfield, Milton Keynes, MK11 3LW, UK
UKHW021256180426
11947UKWH00011B/810